Sex on the Brain

Other books by Deborah Blum

The Monkey Wars

Coeditor, *A Field Guide for Science Writers*

DEBORAH BLUM

•

Sex on the Brain

The Biological Differences Between Men and Women

•

VIKING

VIKING
Published by the Penguin Group
Penguin Putnam Inc., 375 Hudson Street,
New York, New York 10014, U.S.A.
Penguin Books Ltd, 27 Wrights Lane,
London W8 5TZ, England
Penguin Books Australia Ltd, Ringwood,
Victoria, Australia
Penguin Books Canada Ltd, 10 Alcorn Avenue,
Toronto, Ontario, Canada M4V 3B2
Penguin Books (N.Z.) Ltd, 182–190 Wairau Road,
Auckland 10, New Zealand

Penguin Books Ltd, Registered Offices:
Harmondsworth, Middlesex, England

First published in 1997 by Viking Penguin,
a member of Penguin Putnam Inc.

10 9 8 7 6 5 4 3 2 1

LIBRARY OF CONGRESS CATALOGING IN PUBLICATION DATA
Blum, Deborah.
Sex on the brain/Deborah Blum.
p. cm.
ISBN 0-670-86888-4 (alk. paper)
1. Sex differences. 2. Sex (Biology) I. Title.
QP81.5.B58 1997
612.6—dc21 96-52034

This book is printed on acid-free paper.
∞

Printed in the United States of America
Set in Adobe Garamond
Designed by Pei Koay

For Lucas and Marcus

ACKNOWLEDGMENTS

This book would not have happened without a great deal of help and support, and it gives me real pleasure to say thank you.

I am indebted to many scientists for taking the time to share their ideas and research with me. In particular, Kim Wallen, of Emory University; Marc Breedlove, of the University of California, Berkeley; and Judy Stamps and Sally Mendoza, of the University of California, Davis, not only provided information, references, and sources, but allowed me to bounce my ideas off them, kept me from traveling astray, and responded always with great thoughtfulness.

Others took much extra time and trouble in providing research papers and answering my questions, my follow-up questions, and then more questions, and I would like to especially thank Ruben and Raquel Gur, of the University of Pennsylvania; Thomas Insel, director of the Yerkes Regional Primate Research Center in Atlanta; Donald Grayson, of the University of Washington at Seattle; David Gubernick of the University of California, Davis; Paul Lombardo, of the University of Virginia; Adrienne Zihlman, of the University of California, Santa Cruz; Alan Booth, of the Pennsylvania State University; Bobbi Low, of the University of Michigan; and Daryl Bem, of Cornell University. I also appreciated the thoughtful discussions of their work by David Amaral, Randell Alexander, Jay Belsky, James Dabbs Jr., Frans de Waal, Marian Diamond, Anne Fernald, Ellen Frank, Dean Hamer, and John Wingfield.

I also am grateful to the researchers who read and improved the manuscript: Michael Bailey, Jay Belsky, Marc Breedlove, Jane Brockmann, James Dabbs Jr., Frans de Waal, Christine Drea, Donald Grayson, David Gubernick, Dean Hamer, Paul Lombardo, Bobbi Low, Sally Mendoza, Ron Nadler, and Kim Wallen. My husband, Peter Haugen, a literature major and theater critic, read and polished the whole manuscript, argued with me, encouraged me, and transformed some of my most convoluted thoughts into clarity. And Viking's talented production staff—especially copy editor Cathy Dexter and production editor Barbara Campo—were enormously helpful.

I am fortunate to work at a regional newspaper, the *Sacramento Bee*, that has an unwavering commitment to science writing. When I became interested in behavioral biology, the editors encouraged me to explore the subject through an in-depth series, "Only Human," which appeared in late 1995. Some of the interviews in the book—for instance, the discussions with Richard Lynn, Steven Petersen, Bruce Perry, Jerome Kagan, and Tom Gordon—are drawn directly from that series. For all their support and encouragement, I would like to thank my editors—Janet Vitt, Bill Enfield, Marjie Lundstrom, Rick Rodriguez—and especially executive editor Gregory Favre, who has always challenged me to do even better.

My friends put up with me while I became obsessed with the subject of sex differences, tried out my theories on them, and asked them all kinds of personal questions—which they answered with good humor and only the occasional roll of the eyes. My special gratitude on this count to Jim Richardson, Chris Bowman, Daryl Metz, Janet Fullwood, and Maria LaPiana.

My agent, Suzanne Gluck, convinced me that I should write a book on gender biology and, as always, stayed through the process as adviser and friend. And I felt blessed in my editor, Dawn Drzal, who made me work harder than I wanted to but made the book a great deal better as well.

Deborah Blum
Sacramento, California
December 1996

CONTENTS

INTRODUCTION

There comes a moment in everyone's life when the opposite sex suddenly appears to be an alien species. Totally and mind-bogglingly different. The world cleaves apart, with "us" on one side and "them" on the other.

I came to this epiphany late because I was raised in one of those university-based, liberal-elite families that politicians like to make fun of. In my childhood, every human being—regardless of gender—was exactly alike, and I mean exactly, barring his or her different opportunities. Even Santa Claus felt strongly about this, apparently. One Christmas, I received a Barbie doll and a baseball glove. Another brought a green enamel stove, which baked tiny cakes by the heat of a lightbulb, and also a set of steel-tipped darts and a competition-quality dartboard.

It wasn't until I became a parent—I should say, a parent of two boys—that I realized I had been fed a line and swallowed it like a sucker (barring the part about opportunities, which I still believe). This dawned on me during my older son's dinosaur phase, which began when he was about two and a half. Oh, he loved dinosaurs, all right, but only the blood-swilling carnivores. Plant-eaters were wimps and losers, and he refused even to wear a T-shirt if it was marred by a picture of a stegosaur. I looked down at him one day as he was snarling around my feet and doing his toddler best to gnaw off my right leg, and I thought, This is not a girl thing—this goes deeper than culture.

Raising children tends to bring on this kind of politically incorrect reaction; another friend came to the same conclusion watching a son determinedly bite his breakfast toast into a pistol that he hoped would eliminate his younger brother. Once you get past the guilt part—Did I do this? Should I have bought him that plastic allosaurus with the oversized teeth?—you end up asking the far more interesting questions that lead into gender biology, where the issues take a different shape: does a love of carnage begin in culture or genetics, and which drives which? Do the gender roles of our culture reflect an underlying biology and, in turn, does the way we behave influence that biology? Can one redirect the other?

The point I'm leading up to—through the example of my son's innocent love of predatory dinosaurs—is actually one of the more straightforward in this debate. Most scientists agree that the gender difference in aggression begins in biology. Behavioral biology—in the modern sense of gene mapping and brain imaging—is a new science, so the details have not been worked out. But at a 1995 conference on crime and genetics, when scientists were asked how much energy went into figuring out male versus female aggression, their reply was that the difference was so accepted—such a given—that it really didn't make for interesting research.

"There's plenty of room in society to influence sex differences," says Marc Breedlove, a behavioral endocrinologist at the University of California, Berkeley, and a pioneer in defining how hormones can help build sexually different nervous systems. "Yes, we're born with predispositions, but it's society that amplifies them, exaggerates them," he continues. "I believe that, except for the sex differences in aggression. Those are too massive to be explained simply by society."

To pursue the question further, let's consider some hard numbers. Aggression, at least as measured by crime statistics, exhibits an enduring gender gap. Government statistics in both the United States and in Europe record between 10 and 15 robberies committed by men for every one by a woman. I cite this first because it shows one of the biggest differences. (For some uncertain reason, robbery exaggerates the violence gap between men and women. Go to simple assault, and the ratio narrows: three men for every one woman. You can argue, of course, that assault may be a matter of rage, indicating loss of control, whereas robbery requires some cold-blooded intimidation. In the latter case, some obvious biology—men are usually larger—might favor the intimidation theory. One study, in fact, found that dedicated robbers tended, as a group, to work out almost obsessively, building muscle bulk.

But guns, you would think, could equalize that. If women are simply handicapped by size, then they should be able to compensate by arming themselves. For a while, scholars argued that guns would equalize the relationships between men and women; that larger size would be rendered irrelevant. That hasn't happened, and the arguments today tend to concern why it hasn't. Women are just not drawn to guns in the same numbers that men are. Among women robbers as a whole—presumably including amateurs or one-time, desperate cases—only 30 percent use a gun. Among men as a whole, the figure is 50 percent.

There's a similar pattern in domestic-partner murders: close to 75 percent of men who kill their partners use a gun. Women kill their partners too, but again—despite the equalization factor—only about half use a gun. And women almost never stalk their previous lover, gun in hand, after a relationship is ended.

There's another telling statistic in the domestic killing studies: In conflicts where a woman killed a man, he tended to be the one who started the fight—in 51.8 percent of the cases, to be exact. When the man was the killer, he again was the likely first aggressor and by an even more dramatic margin. In fights in which they died, women started the argument only 12.5 percent of the time.

Asking why and how our species developed such a strong gender difference tends to lead us into the world of evolutionary psychology, which tries to reconstruct the way our past shaped our present behavior. There's an evolutionary theory—some would call it influential, even—predicting that (overall) men will flash more quickly to anger and aggression; that women will be kinder and calmer. The theory is based on the idea of minimum investment in the survival of one's own genes: What's the least a male or female has to do to produce a child and, therefore, genetic continuance?

Stripped down like this, the contrast is extraordinarily sharp. At minimum, a male must have healthy sperm and an interest in sex. At minimum, for mammals like us, a female must produce an egg, carry and provide nutrition for the developing fetus, and, odds are, take on much of the child care. As we know only too well, she does not have to be interested in sex to get pregnant.

Thus, if we think of behavior in terms of evolution—the survival of the species, and of our own particular genes—the difference becomes compelling. How could the same strategies suit both sexes? Healthy, fertile females don't, in general, have to fight to get pregnant. Their chances of reproducing, of continuing their genetic line, are very high. Some would say 100 percent. But males do often have to fight to become a parent—which is the central point of this evolutionary argument. Over the thousands, even millions of years, there's been enormous evolutionary pressure on males to be the ones who hustle, who push, who have power, who achieve the prize of sexual access. In monogamous species, where a long-time mating arrangement happens, the pressure is less. But among polygamous species, competition may be everything and may be brutal. Most of the world's species are polygamous; many scientists argue that early human ancestors were as well.

One of the most dramatic examples of how desperate this battle for genetic immortality can be comes from elephant seals, where the clearly polygamous society is dominated by huge bull males who claim most of the females. Biologists calculate that only 20 percent of the males ever become fathers. And there are many similar examples—orangutans live in such an ironclad harem system that for many young males, their only chance to father a child is through rape.

Obviously, in such a scenario, pushy, even violent males, quick to take on challenges, might well stand the best chance of reproducing. Natural selection might have favored those males who would risk it all for immortality. But where, in evolution, is the comparable incentive for females to become so physically nasty over sex? They seem surrounded by potential and even overeager mates. It makes more sense for the female to be choosy—pick a mate carefully, go for the good genes.

Further—and remember, this is the stuff of basic instinct and evolutionary background—fighting doesn't seem to give females a strong advantage in terms of carrying genes into the next generation. Early man might have battled his way to several mates and thus several chances of reproduction, but in our species, a woman doesn't have more than one successful pregnancy a year. So, in terms of reproduction, risking it all for multiple one-night stands might make sense for men, but not women. In reproductive terms, she'd do better to save her energy toward mothering the baby into adulthood. Her greatest risk of failure is in the death of a child. By such evolutionary theory, we should clearly expect men and women, directed toward such separate goals, to behave differently. We might also expect what we do see today—more aggressive males.

It may seem weird to be talking about elephant seal sex and early human parenting at this point. But the basic theory—and it is basic to the extreme—is fundamental to the way biology explains sex differences. Many scientists trace almost every behavioral gender difference—from emotions to competitiveness—back to the different demands of being a sperm-producer or an egg-maker. The path is not always direct, but it winds in a curiously logical way.

This is a very broad concept, relating only to general behavioral influences. It doesn't explain the specifics of becoming a serial killer, say, or an international banker. It doesn't separate humans from other species, either. Evolutionary psychology is not just about humans; it pulls us into the well-rounded world of Earth-based biology. It says that we share some behavioral urges with other species; the same kind of reproductive argument would suggest that bulls would be more aggressive than cows, that male dogs would chase females, and not vice versa. The argument does not rule out females being competitive or even dangerous, especially when it comes to survival of the young (think of the risks involved in stepping between a mother grizzly bear and her cubs). The point is that these influences may run through our own primitive evolutionary past, that they remain with us in subtle ways, and that perhaps they lay the foundations of today's culture.

How do you test that, though, and how do you distinguish cultural from biological influences anyway? University of Washington archeologist Donald Grayson tried this a few years ago, after also reaching an epiphany of sorts. Grayson's specialty is the ancient cultures of the American West, particularly those built in the deserts. So he's spent many field trips in some of the least hospitable lands on Earth, sandblasted by the wind and basted by the sun. (And, of course, there's always the occasional disaster of equipment failure and overheated tempers.) What Grayson found was that come a crisis—say, a Jeep with a bad engine, stranding his group in the sandy heat—it was the men who jumped to fix it. They were efficient and cheerful as long as they could make the repair, and make it quickly. But if they couldn't, if the crisis began to drag into a lost day, then the men—himself included, Grayson acknowledges—became grouchy and irritable. They were depressed about the situation and increasingly frustrated about being stuck with a bunch of people they suddenly didn't like too much.

"At that point, women were the glue," Grayson said. Women comforted and teased, proposed other solutions, and kept the group from igniting in a blaze of frustration and irritation. Again, some of this has to be attributed to culture—don't we teach women to be nice and men to be fixers?—but it was so consistent and so enduring and so effortless that Grayson began to suspect it ran deeper. He began to think about the different biology of men and women in terms of strengths.

And he wondered, if you examined men and women in a time of total disaster—of life-threatening terror—when people often are stripped down to basics, would the same pattern emerge? And would it make a difference? Would the reactions and capabilities of one sex serve better than another? He decided to look at one of the most famous, or notorious, tragedies of the early American West: the awful fate of the George Donner Party of 1846, in which a group of pioneers were trapped by snow in California's Sierra Nevada mountain range.

The Donner Party consisted of a group of 87 Midwesterners, mostly farmers, lured by tales of golden California. They were brave; they were naive; they knew almost nothing about the wilderness before them. Along the trail—and their route often led through canyons with no trail at all—they fell too far behind schedule. Winter literally roared over them; the snow banks around their camp reached a prison-wall height of 22 feet. Almost half the members of the Donner Party died in the mountains. But the group is best known because those who survived did so by eating those who did not—stewing the bodies of men, women, and children.

Would you call this life on the edge? Grayson did. In that framework, he found a number of startlingly consistent gender differences. Female survival stood out: 57 percent of the men died compared to 30 percent of the women. Why? First, because the men killed each other; three of the 40 deaths in the Donner Party were murders. Second, because the women's bodies were better built to handle cold and starvation. Women have more insulating fat, and their bodies are, on average, 17 percent smaller than men's and burn up less energy. And, third, by the time total crisis struck, the men were far more worn out. Bigger and stronger, they were the ones who hacked paths for wagons to follow, carried exhausted children, built shelters. It's logical for a society to assign the heavy lifting jobs to those with the largest muscles, of course. But when starvation came, the men had fewer reserves than the women.

Grayson also suspects that men are less durable, period. Returning to the issue of reproductive differences, he says, there's good reason for nature to design females—the primary reproducers of a species—to last a long time. Women outlive men by an average of seven years these days, and there's new evidence that hormonal concentrations in women effectively prolong good health.

In looking at longevity, Grayson also analyzed another pioneer disaster, the Willie Handcart Party's crossing of the Rocky Mountains in 1856. The party belonged to an experimental Mormon program in which emigrants pushed their belongings in handcarts rather than riding in the slower, more expensive wagons. However, the 1856 party, led by James Willie, ran into winter snows. They were trapped, exhausted and short of food. Of the 429 people, 68 died on the path between Iowa and Utah. And although men and women were equally tired—everyone walked, everyone pushed the carts—the dead numbered 40 men and 28 women. (The percentage breakdown sounds even more dramatic: 25 percent of the men, 8.5 percent of the women—a three-to-one ratio.)

There's also another contributing factor which may highlight gender differences. The Donner Party was stranded because the men who led it decided to gamble on the best way to reach California. They chose an untested route, hoping it would be quicker. They left the well-known trail. There's evidence that they did this over the objections of the women in the party. And certainly the evolutionary argument that I've just outlined would predict that split: that men would be far more inclined to take the chance.

Stephen McCurdy, a professor of occupational medicine at the University of California, Davis, did another gender-focused analysis of the Donner deaths. His conclusions agreed with Grayson's. "In the decision to take that shortcut that led to destruction, the women were not involved," McCurdy said. "I hate to throw a monkey wrench at my own sex, but I can't imagine a woman being willing to take that risk with her own children and family."

If, then, women come less easily to risk-taking, they also move less readily to destruction and harm, at both high and low levels. Jay Belsky, Distinguished Professor of Human Development at the Pennsylvania State University, talks about risk-taking on a small scale. "Dads like to toss their babies in the air," he says. "And I guarantee you, wait long enough and the tossing gets higher, and pretty soon, the wife is saying, 'Honey, that's enough.' She'll stop him. If you look at emergency room statistics I suspect you'd see more children injured on Dad's watch than on Mom's. Dads don't watch in the same way; they're more willing to take risks, even with their children."

Grayson also believes that men are simply more naturally inclined to gamble with lives. All of these studies tell us, Grayson thinks, that history is really the story of biology.

Of course, you can't prove any of this unless you can figure out the mechanics—what "biology" are we talking about specifically? Do our genes have a program in mind for each sex? Do they produce a distinct male chemistry that leaps to the fight and a distinct female one that turns toward comfort and caution? It's undoubtedly not quite that simple, but scientists studying such questions propose hormones as the most likely candidates in this scenario. For instance, males run high in testosterone, a hormone that spikes up in times of challenge. Perhaps—and this remains speculative—the hormone also propels a quick, aggressive response. Females are high in oxytocin, a hormone that supports not only lactation but, apparently, bonding and nurturing behaviors—although this too remains speculative.

This is not to suggest that we dangle like puppets on hormonal strings. If hormones do profoundly affect behavior—which I believe, though not all do—then they must do so as one of many cast members, not as a solo performer. Our behaviors are, in many ways, wide open to many influences—foods, drugs, injuries, life in all its dimensions. We can also choose to override an instinct. Think of biology and behavior as dancers—one leads, the other follows. But which does which, and when? They tug at each other, and in turn are pulled by the music, the fluid melody of the environment. And do we ever know—can we ever know—where we are in the dance?

It's that sense of motion which makes gender biology so frustrating, so promising, and so blessedly interesting. We're trying to figure out who we are, even as we're changing into someone else, moving from the waltz to the two-step. And because we do change, there's no limit to the questions we can ask, from the grandiose ones (Why do we have two sexes at all?) to the small ones (Is there a scientific reason why boys get intense about Legos and girls—as the manufacturer's studies show—do not?). How easy is that to change? Do we want to?

Late in the summer of 1996, the *New Yorker* magazine ran a cartoon showing a man sitting at a kitchen table, relaxing with a newspaper. His wife, children dangling off her arms, was setting dinner down on the table. She was looking at him without affection. The caption said, "Despite my best efforts, you're still the man and I'm still the woman."

In that image is the bottom-line question, the one that I want to pursue in this exploration of gender biology: Are men and women dancing toward a real and equal partnership and is it biology that makes it, apparently, such a very slow dance?

•

Chapter One

SPLITTING THE LARK

The Origins of Sex

•

In the realm of poet and playwright, the coming together of the sexes is all beauty, all passion. Think of Shakespeare's Romeo, who stands outside the house of a girl he's barely met, caught by an unexpected enchantment. When she appears at the window, it's like dawn's first shimmer on the horizon: "It is the east, and Juliet is the sun."

There's no metaphor too great or too gorgeous. Man and woman are restless tide and welcoming shore. All hope is in a face, all fortune in a gilded body. The whole affair becomes so exhilarating, and ultimately so exhausting, that even the flamboyant nineteenth century romantic, George Gordon, Lord Byron, acknowledged it could go too far:

For the sword outwears its sheath,
And the soul wears out the breast,
And the heart must pause to breathe,
And Love itself have rest.

In the literal world of science, of course, Byron couldn't have been more wrong. Lungs breathe. The heart pumps. If it pauses, call the doctor. "Love itself" can be broken down to hormonal signals and molecular structure. This conflict—the ruthless exchange of shining beauty for clinical chemistry—has endlessly frustrated poets. More than 100 years ago, Emily Dickinson wrote sarcastically of dissecting birds to find the mechanics of song. "Split the Lark—and you'll find the Music—" she said; kill the bird and silence the melody: "Scarlet Experiment! Sceptic Thomas! / Now, do you doubt that your Bird was true?"

In Tom Stoppard's recent play *Arcadia*, which turns physics into an exploration of life and love, a literature professor rails against a world pulled down to the rattle of atoms and the empty clink of test tubes. He takes up Byron's lush verse like a weapon, reciting:

She walks in beauty, like the night
Of cloudless climes and starry skies;
And all that's best of dark and bright
Meet in her aspect and her eyes:

Let science give that much glory to life, then; let research enrich our lives by half that measure—and the professor slams off the stage.

Even scientists sometimes wonder if too much analysis misses the point. At a late-1980s conference on the evolution of sex, the famed biologist John Maynard Smith confessed to a gut reaction against explaining away sexual attraction as nothing more than a means of fixing genetic errors: "I find it hard to accept that bird song, the color of flowers, *Romeo and Juliet*, all evolved because of a molecular mechanism of double-strand DNA repair (and nothing more). Somehow this explanation seems inadequate to explain the phenomenon."

Yet, he continued, "We should not rely on gut feelings in science." Research will never transform women into starry skies and morning light.

Science offers a different kind of beauty. It opens the worlds within worlds. We can hear not only the bird's melody but his message as well. A lark's song becomes more than aimlessly pretty notes; it's the sound of seduction, the bird world's version of a pick-up line—although perhaps more charming than most of ours.

John Maynard Smith is right. Science does—or at least it tries to—strip sex down to the fundamentals of biology. I'm not arguing for chemical romance. I'm not saying that a digital computerized image of infrared radiation stirs the senses the way Byron does. Neither can I say that current explanations for human partnership—for instance, that, like mice, we may sniff out our mates—exactly warm my heart. Poets and dreamers—and the rest of us, too—enjoy love's mystery, an inexplicable alchemy that turns lead to gold, binds heart to heart. What science substitutes, or tries to, is a promise of understanding, a chance to see ourselves through another lens.

When it comes to sex, that vision doesn't come easily. Biologists are still trying to figure out why life demands two sexes. From the point of view of simple, clean reproductive biology, this seems unnecessary duplication. If species need only reproduce themselves, as generously and prolifically as possible, then the logical answer seems to be self-contained clone generators. A planet inhabited by hermaphrodites seems more sensible than the messy system we have now, with all its competitions and battles and emotional furors. From the beginning of evolutionary theory, biologists have been stumbling over this one. Charles Darwin professed himself frankly baffled—"The whole subject is hidden in darkness"—and today, at best, it sits in an uneasy twilight. Never mind cloudless climes, rising suns, and breathless hearts—the two-sex system is too incredibly, unbelievably inefficient.

How inefficient? David Crews, an evolutionary biologist at the University of Texas, once looked at that question by analyzing chicken reproduction. (This is a perfect example, by the way, of the distance between poetry and science. Who writes poems about the merits of hermaphroditic hens? Who would want to?) In Crew's scenario—purely hypothetical—a poultry farmer owns ten breeding females. Each hen produces 30 eggs a year. The female chicks begin busily laying eggs themselves when they are three years old. To analyze efficiency, Crews then compared the eventual offspring of a half-male, half-female poultry population with that of an imaginary all-female, self-fertilizing group. The existence of males—incapable of laying even one lousy egg—didn't cut the offspring merely by half. In seven years, if none of the chickens died (and remember that males seem to die off faster anyway), the fortunate farmer who owned the self-sufficient females would own 633,100 egg-laying hens. His competitor, relying on the traditional two-sex system, would own 56,150 hens.

Furthermore, if counting unhatched chicks seems figuratively risky, mating is literally so. In the predatory wild, sex demands a relaxing of vigilance. There's a concentration requirement—so to speak—that interferes with the normal alarm systems; the senses aren't tuned to the pad of heavy paws in nearby bushes. For wild animals, the act of mating can be a flirtation with death. You can measure this, in a way, by realizing how rapidly most animals get through it, as if they really want to get back to looking over their shoulders. Among the apes, the act of copulation has a breathtaking sense of hurry, averaging seven seconds for common chimps, 15 seconds for bonobos, sometimes known as pygmy chimps, and one minute for gorillas, who, after all, are big enough that most predators aren't going to mess with them.

Scientists can't hold that kind of stopwatch on their fellow humans. It makes for a hilarious image: Okay, 30 seconds; okay, 60 seconds . . . As you might imagine, this activity ends up being estimated instead by means of a nice, safe survey. And by such self-reports, we find that the American average for length of copulation time is about four minutes, starting from the moment of penetration. You have to wonder about this number, though. Is anyone going to admit they have sex like a gorilla? Do people feel a competitive urge to report a better time than their neighbor's? The four-minute mark is best taken for its main contrast with the habits of other primates: humans take longer. Perhaps the many years we've lived out of the wild has allowed us to relax the pace. Still, there remain other risks to copulation. We could have spared ourselves sexually transmitted diseases like syphilis, chlamydia, and AIDS, if we had evolved as clean, self-reproducing hermaphrodites.

There are similar risks for other species. Yet it's a male-female world: the birds, the bees, the butterflies, the fish, the frogs, the trees, the flowers, the bushes—on this planet, if it breathes, it almost always belongs to one sex or the other. Two sexes are a fact of life. Why even ask why? Why not concentrate instead on how to live with it?

Maybe it's worth asking only because sex is so fundamental to life on earth, to who we are now. We know men and women are different. We bemoan the differences, sometimes jokingly, sometimes with real dismay. Perhaps our best chance, though, of narrowing the gap is to understand it. The sexes are unlike each other because humanity—as well as the vast majority of species—evolved that way. Earth-based biology likes a two-sex system. Evolution has conserved it, probably for billions of years, because it works. David Crews's hypothetical example of hermaphroditic chickens—as he freely admits—works beautifully as long as there is no need to worry about lethal genetic mutations or the other vulnerabilities of an inbred system. That's why the superefficient chickens remain a hypothesis and, biologists think, why self-replication is incredibly rare. It doesn't offer the survival advantages of genetic variability. If biologists are right, a one-sex system would have doomed us all long ago. Earth would be as silent as its dusty moon.

The reasons for sexuality track the history of life on this planet. The rules of survival begin in the blast-furnace origins of the solar system, in those long-ago days when the sun was settling its satellites into place. Consider the following, then, to be a hyperfast trip from 4.5 billion years ago, when the planets were smoking, to the present.

Scientists suspect that sex was, in the beginning, a primitive means of staying alive. If the Earth seems an unfriendly place now, consider the early planet—a ball of barely cooled rock flung into orbit by a still-sparking star. The primitive atmosphere was hardly the cushy, oxygen-dense one we rely on now. Many scientists believe that the swirling, breath-giving levels of oxygen that we inhale so automatically now are a by-product of plant life.

Plants "breathe" differently than we do—in just the opposite way, in fact. They take in carbon dioxide and release oxygen. The more plants the earth supported, the more oxygen was pumped into the atmosphere. And oxygen, fortunately for us, is a marvelous element. It combines with hydrogen to make water. Left alone, it bonds with itself, in pairs and in triplets. We breathe the former —the oxygen in our air. But the latter, otherwise known as ozone, forms an atmospheric buffer—the famous ozone layer that protects us against the more lethal aspects of living close to a star. Ozone blocks ultraviolet (UV) radiation, a light invisible to the eye and deadly despite that.

In those first days without much oxygen, there wasn't much ozone layer. The atmosphere, by best guess, was a flimsy blanket of other gases—a lot of nitrogen and hydrogen. The earth baked in the best and worst of the sun's glow, heated by infrared radiation, brightened by visible light, blistered by ultraviolet light. That tale of wispy atmosphere and boiling solar wind, believe it or not, is thought to be the beginning of sex. As life tried to emerge—tiny slimy bacteria-type life, but still life—sunlight played a undeniable part. It was not in every respect a healthy relationship.

It turns out that the ultraviolet component of the sun's rays, whether by coincidence or not, burns at a wavelength that interacts dangerously with Earth-based biology. It fits like a key into the locks of DNA—deoxyribonucleic acid, the carrier of genetic codes. Almost casually, it flicks the tumblers of the genetic material within. In comparatively large and complex creatures like ourselves, the relatively small amounts of UV light that get through can still bring on skin cancer or induce the formation of cataracts in the eyes. On ancient Earth, with its wide-open atmosphere, it could have blown a fledgling bacterium apart.

Bear with me; we're taking a sudden, breathtaking leap across the evolution of life itself and the still mysterious appearance of water. But the main point is that when primitive organisms began to create squat little pads and clumps in the newly formed seas, they were hammered by light. During this period, from about 3.8 billion to 2.5 billion years ago, life just held on. Survival itself was a triumph. And biologists think those clumps of one-celled creatures survived only because they somehow developed a system of transferring genes. If one genetic combination was vulnerable to radiation, perhaps another was less so. If two organisms—perhaps bacteria—could mix their genes, each stood a chance of replacing vulnerable ones, finding stronger ones, developing better defenses. The early gene transfer essentially involved bacteria melting into each other, with one sucking up the other. But it was also the union of two different genetic packages: the theoretical beginning of sex.

And since life does tend toward efficiency whenever possible, two "parents" turned out to be adequate—as opposed to a reproduction system involving, say, 21 different partners.

As far as sex goes, biology's rule, at least for reproduction, is that there must be two. It's as true for spiders as people, for banana plants as hummingbirds, fish as hyenas. Anyone with the energy can have countless sexual partners, but to reproduce requires only one—and good timing.

That's not to say that the system hasn't gotten more complicated. True, there are still bacteria that use the merge-and-suck system, but that's because nature hates to throw out what works. Moving on up the food chain, most of us—from bugs to birds, fish to humans—have moved past the genetic vacuum cleaner phase. Instead, we bring different genes together through a process of cell fusion—or fertilization. This is the bringing of the pollen to the plant, the sperm to the egg. As those cells meet and fuse—in our case, when tiny tadpolelike sperm enter the rounded blob of the egg cell—we have the start of a new life. Once again, we can ask what started this new system of sexual exchange. Once again, there's no romance in it whatsoever.

One theory is cannibalism. As you might imagine, the early earth was a barren place. First life was desperately seeking not only genes but food. Say you're a tiny, simple cellular blob, surrounded by hot rock and murky water. Realistically, what are you going to eat? Gnawing on rock is not an option. Single-celled creatures, so scientists speculate, probably devoured each other. They could have forged a tightly woven food chain in which cells were constantly merging and sharing, blending their components.

Once the system really got moving, of course, it was less like cannibalism. The recombination of genes led to different forms of life. But once there were algae and eventually larger plants pumping out oxygen, why did those diversifying life forms stick with dual sex? After all, ultraviolet radiation was finally being buffered and the food supply was expanding. You might think that once the system was running, it might relax into single-sex simplicity. As if life were ever that simple.

Environment is a lot more than sun, rock, and water, after all. It includes other life—some oblivious, some useful, some downright vicious. Researchers suspect natural conflict kept the sexual pressure on. For every creature that might wish to be genetically lazy, there is another keeping life from getting too comfortable. The chain of life includes predators and prey, parasites and unwilling hosts, viruses and victims. If you look at predatory cellular parasites or viruses, they often resemble professional change artists. Some change their outer proteins—the chemical identity by which a host organism might detect them and defend against them—almost as easily as we change from business suits to sweats.

The retrovirus family, which includes the Human Immunodeficiency Virus (HIV), is forever shrugging off yesterday's coating. The influenza viruses mutate into new infections every year. All of that puts new demands on the hosts to be equally shifty in their defenses. In other words, both need an ongoing genetic recombination. Mating offers a quick and constant way to reshuffle the genetic deck.

We can use a species of freshwater African snail to illustrate the point. Scientists at Emory University have found that the snails use sex not because they can't reproduce without it—they do just fine in that respect—but because sexual recombination of genes is their response to parasitic attack. The attacker is a tiny thread of a worm—scientists call it *Bulinas truncatas*—also known for weaving

its destructive way into human bodies, destroying livers and kidneys.

The shiny little snails prefer to live as self-contained hermaphrodites, and in the dry season in Kenya, when the parasite load is low, each snail stays contentedly in its self-contained shell of a world. But as the rains pick up, and as the parasites spread viciously in the steamy damp, the snail needs more genetic flexibility to protect itself than an asexual approach provides. Remarkably, the snail is able to recognize this. It changes under this pressure: part of the population begins growing penises, engaging in sexual intercourse with the rest, and shuffling genes to find whatever combination will best outlast the parasite this time around. It's their offspring who really benefit from this, developing a tougher resistance when necessary.

Computer simulations comparing asexual, or hermaphroditic, life, which recycles the same genes over and over, with a sexual system suggest that asexual life usually self-destructs. If a bad genetic mutation occurs, it can't be shaken off. Barring a miraculous countermutation, the organism can't substitute a better gene. University of Oregon analysts running such a computer-based experiment said that introducing parasites into such a fragile biology pushed their asexual computer creatures right into extinction.

The biological objective of sex, then, is genetic flexibility. It's mandatory; if the environment changes, life must adapt to it or be killed by it. Consider the ozone layer—and we probably should, considering our own recent efforts to repair damage we've done to it. In general, biologists believe that when the ozone layer formed long ago, and as ultraviolet radiation was better screened, Earth's first creatures took the opportunity to grow and spread and compete with each other. Life itself became part of the changing and challenging environment. There were new kinds of pressures on genes: those often-deadly dances between predator and prey, virus and victim. The pressure continues. We match it with a remarkably determined two-sex system.

Consider another African species, the cichlid fish, which lives in the shallow waters of Lake Tanganyika. The cichlid are organized in a strict social system in which ten percent of the males are dominant and boss the others around, controlling the food-rich, sunny waters that attract spawning females. These boss fish are bright-colored and virile. The rest of the males are pale and sexually immature.

But if one of the big guys gets scooped up by a hungry bird, suddenly a new boss appears. Since there's an ingrained biological insistence on having a certain ratio of virile males to fertile females, one of the shadowy male wallflowers is suddenly transformed. Stanford University scientist Russell Fernald calls it the switch from wimp to macho. The new boss fish develops black stripes and orange spots—gaudy, to be sure, but apparently the right color scheme to bring on the female cichlid. His sex organs mature; over a period of days, he grows to eight times his former size. He's tailored to mate.

For most species, then, the first requirement for producing offspring is two parents of the opposite sex and, as the cichlid transformations show, some species do extraordinary work to get there. The second requirement, at least for having healthy children, is that the notion of "opposite" be taken very seriously. Obviously, you'd expect parents of different sexes to merge different genes in producing young. But (within reason) the more different the better. The less alike a couple's genes are, the more variability, the greater the ability to respond to genetic injury, and the less likely that defective genes will be inherited. As we know, the consequences of too much genetic duplication can be tragic. Consider this hypothetical example: A woman marries a second cousin. They find out later that their family line carries the gene for cystic fibrosis. As a rule, two defective genes must be inherited for such a disease to lock in. Their children, obviously, are far more at risk than if the parents had married into other families with different genetic histories.

Biology operates on the guiding principle that different is better. This drive is strong enough that some biologists even argue that sex itself is a factor in genetic instability, in mutations which—for better and for worse—translate into "different." Ursula Goodenough, at Washington University in St. Louis, even suggests that we should expand our notion of sexual influence beyond the obvious. Goodenough believes that genes which control the immune system, for example, or that help guide our sense of self or nonself may also be an important part of the two-sex system and may subtly push us toward choosing mates with very different genes from our own. This reinforces the notion that the two-sex system is geared toward genetic diversity.

Goodenough's primary work has been done with a one-celled green algae, chlamydomas. Most of us would identify this as slime; she sees a family. One distinct group of the algae oozes along the east coast of the United States; another equally mucky branch hangs off Cuba. What she's found by comparing the American species (*reinhardtii*) and the Cuban species (*incerta*) is that the greatest genetic differences are in genes related to sex and reproduction rather than those that control digestion, say, or motion.

Goodenough began by wondering why there should be two separate groups of algae at all. The North American species dwells right down to the coast of southern Florida, just a narrow ocean current away from Cuba. The environmental niches are hardly different enough to split into separate species. Goodenough's argument is that it isn't environment—or that wedge of water—that separates these two. Rather, she believes that genes related to sex are programmed to be more unstable than others to encourage variation and difference. If the sexes indeed evolved to keep the gene pool in motion, they might have also evolved to be prone to mutation.

Her fascination with the sexual drive toward differences led her to consider the importance of immunology in courtship. She's begun to look at what she calls the "self-incompatibility" genes, which are involved in creating the body's sentry system, such as cells in the blood that recognize foreign objects. Self and nonself recognition is an essential part of the immune system, allowing the body to recognize invading bacteria or viruses as alien, or dangerous. But, as Goodenough points out, the system also may be part of healthy reproduction. It may guard against inbreeding, making sure that one organism doesn't mate with another that is too close genetically. Flowering plants, for instance, have an elaborate self-recognition system that will abort the interaction of a male pollen cell with a female (stigma) cell if the pollen seems too closely related to the "self"—or female cell. It's a measure of their importance that such rules of avoidance seem deeply grained in genetics. The broccoli plant, Goodenough points out, has 50 different kinds of genes for avoiding "mating" with a too-similar broccoli.

What she suggests is that the immune system guards against both the health problems posed by alien invaders, such as viruses, and those posed by sex with a too-familiar party. The demands of such protection may keep pressure on the genes to keep up. "That is, [these genes] evolved rapidly to generate immunity to inbreeding," Goodenough says. If she's right, the immune system is partly sexual, a part of mate choice. Is she right? She doesn't know. What she and her colleagues are doing is asking questions about how humans choose each other. Are there influential chemical signals and scents? Is information about internal chemistry, about the immune system, conveyed in such subtle ways? Do our genes thus, quietly, try to pick our partners for us? It's important to emphasize, as Goodenough does, this is a science of interesting theories rather than conclusions. If she's right, though, our bodies may well be turned, albeit unconsciously, to some very clear rules of sexual attraction. "Not only must you find a mate within your own species, of the opposite sex, but you must determine if the mate is like you or different from you," she says. The "within a species" constraint is worth emphasizing, of course. A two-sex system doesn't encourage cross-species sex; we look for variation within our own.

In our own species, there's strong suggestion of a biological barrier to like-like mating. Most of us feel no sexual pull toward a brother or sister. And even if we do feel such a pull, we are restrained by social mores. Rules against incest stretch across cultures and time. In early Christianity, marrying a thirty-second cousin was considered incest. The medieval Church of England published a "Table of Affinity and Kindred," which listed some thirty-odd forbidden marriages, including a warning against a woman moving in with her grandfather or a man developing a relationship with his grandmother. In traditional China, a man was forbidden to marry any woman with the same surname. Human societies have regarded incest as so evil, so likely to call down the wrath of the gods, that it has been blamed for catastrophes ranging from crop failures in Ireland to cattle sterility in Thebes.

Of course, the existence of laws also tells us that even if there is a biological predisposition to avoid incest, humans may choose to ignore it. Fathers do rape their daughters; sisters do sleep with brothers. The laws and the penalties tell us two things: that the majority of people regard this as wrong, and that a minority do it anyway. Incest taboos have been deliberately lifted on occasion: In the ancient Egyptian dynasties, the ruling families deliberately inbred to keep the aristocracy pure. Cleopatra was both her husband's niece and his sister. The ruling families of Hawaii, the great Inca rulers of long-ago Peru, the past Chinese dynasties—groups with enough power and enough social snobbery—have chosen incest over sexual commingling with the lower classes. It's worth noting, however, that all the above-mentioned families have vanished.

Is there a biology that underlies the general cultural abhorrence against incest—the almost immediate feeling of revulsion? You could argue that society simply teaches us to regard the idea as horribly wrong. But there's a growing body of evidence to support Goodenough's point about self-recognition and sexual choice—that there's a subtle but protective chemistry beneath it. Incest does, in fact, serve as a reminder: we share a common biology with the rest of life on earth. Even broccoli avoid incest. But we have an apparently unique ability to override biology, ignore predisposition, to choose to hunt down the worst possible genetic partner, often for the worst of reasons.

As Goodenough suggests, research is drawing a tighter link between immunology and sex. Some really fascinating work along this line has been conducted in Switzerland by Claus Wedekind, a zoologist at Bern University. Wedekind began with the obvious functions of self/nonself recognition—basic defense against disease. A body that doesn't recognize a virus as foreign is a body that is open to infection. Among the genes that control our response to not-us-ness is a group known as the major histocompatibility complex (MHC). These genes code for disease detectors in the immune system—not the killer cells, which destroy invaders, but the sentry cells, which patrol the body, sending out an alert if unknown organisms are detected. Based largely on Wedekind's pioneering work, the MHC genes are now under intense scrutiny. The first international MHC and behavior conference was held in July 1996.

Wedekind's early research showed that when a female mouse is offered a choice between two males, she always chooses the one whose MHC genes are least like her own, within the boundaries of the species. She checks the male out by sniffing its urine, in which his immune system is signaling wildly by creating proteins unique to each set of MHC genes. These in turn perfume the urine. The stranger the smell—at least to the mouse in question—the more endearing the male mouse. But Wedekind has also made an unusually neat transition to human behavior, with a logical shift to the way we smell each other. Most of us tend not to sniff each other's pee, so he studied sweat, which is sometimes hard not to smell.

Wedekind collected a group of 44 college women and 49 college men. He gave each man a clean cotton T-shirt and asked him to sleep in it over a weekend. The men were also told to avoid spicy food, deodorants, cologne, perfumed soap, smoking, drinking, and sex. During the day, their sweaty T-shirts were kept in sealed plastic containers and, after the weekend, they were carefully packed back into their boxes and returned to Wedekind's laboratory.

Now, here's the mouse connection. At around the time of each woman's ovulation—during which studies show that sense of smell becomes more acute—the woman was seated alone in a room and presented with a stack of plastic boxes containing both the sweaty T-shirts and some immaculately clean ones. She had to sniff each one. The women were asked to rate every shirt for sexiness, pleasantness, and, I suppose, basic reek, although the scientists called it intensity of smell. The women reacted a lot like the female mice: the more different the man's MHC complex was from the woman's own, the sexier the shirt—and presumably the man—was rated. Although they haven't tested it, the researchers suspect the arrow points both ways; logic says men, too, would get turned on by the aroma of "alien" immunology.

As in mice, as in broccoli, as with roses perfuming the air of spring, the point of having two parents is to combine dissimilar sets of genes. Again, how do we recognize a genetically optimum partner? Again, we don't know all the ways. Wedekind's work suggests one—that the human immune system signals its individual chemistry, as loudly as possible, through compounds released in sweat. That chemistry derives from the genes that control the system, in this case the MHC complex. And women respond to the "information" in the smell. Wedekind's study suggests that women are most attracted to the scent of men with different MHC genes from their own, implying at least that biology disposes an individual toward a mate who would provide a healthy mixture of genes.

Is that the only cue we use? Undoubtedly not. Appearance, for instance, can also provide information about a healthy mate and I'll discuss that more later. But the relationship between MHC genes, smell, and sex is particularly interesting because its power is so subtle—influencing us without our awareness. It emphasizes Goodenough's point about how potent genes may be in guiding sexual choice. As I said, we can also override predisposition. But there seems to be a genetic cost to that. Researchers, for example, looked at the Hutterite religious community of North Dakota, where people marry within a small, confined society. Anthropologists found that those couples with closely matched MHC complexes tended to achieve fewer pregnancies and suffered more miscarriages. Perhaps the signals in sweat are worth responding to, even unconsciously. One of the ideas raised at the July 1996 international MHC conference is that the same immune signal may exist in human saliva. That might mean that kissing—a biological mystery if there ever was one—functions as a kind of mutual genetic test. On that note, I acknowledge that all those disgruntled poets have a point—science really can take the romance out of anything.

But the more essential point of all this—from early Earth to sweaty T-shirts—is that genetic variation is essential. Thus, we are not overrun by a proliferation of hermaphroditic egg-producing chickens. A slight fraction of one percent of all species on earth are hermaphroditic. They tend mostly to be slimy little things—blue-green algae, salamanders, frogs, minnows. Almost all, apparently, were once typical two-sex species that, under unclear evolutionary pressures, basically pulled it together.

David Crews has studied a family of lizards, the whiptails, that includes such unusual members. One third of the 45 species in this group are self-reproducing females. Are they weird as well as rare? Crews has found that they engage in what we can call pretend sex, mounting each other and humping away regardless of the futility of it all. His experiments indicate that this pseudosex actually accomplishes something, stimulating ovarian growth just as male courtship stimulates ovarian growth in other whiptails. The self-reproducing females are still haunted by the ghosts of the ancient male-female dance.

There are also fish, such as the Japanese goby, that flick back and forth between male and female, based on social cues. For the goby, size is power; the biggest goby in any given group becomes both seducer and soldier. He's the first-line defender of the territory. He's the hunk, drawing the attention of all surrounding females. It helps to be the group's only male, of course. But this poses a dilemma as well: What happens if a passing bird spikes your only male? It may be lunch to the bird, but, at first glance, it's catastrophe for the gobies—no male-female mix, no mating, and no future.

The saving grace is that sexual biology is remarkably responsive to environment. Biologists have found that the goby can keep this social system because it can also switch sexes. If the king male is gone, the largest female goby changes into a male. This is so flexible a system that researchers can yank the gobies back and forth. They put a group in a tank and remove the dominant male. A female converts. They drop the old leader back in. Sure enough, the erstwhile boss reverts back to female, with all her original enthusiasm for the male.

There are also species, such as sheephead fish, crocodiles, and lizards, for which the environment factors into whether their offspring will be male or female. How warm the mother's internal body temperature gets during incubation, for instance, can determine whether the eggs hatch as male or female. Take the leopard gecko, for instance—a gorgeous little species of lizard, with its flashy, jungle-foliage markings. Temperature extremes tend to produce gecko females; medium-range temperatures make males. The result is about what you'd expect from the chance randomness of weather: a 50-50 male-female split, close to ideal in the gecko social system.

Would any of those systems work for us? The gender-on-demand system of the goby? The sex-by-weather responsiveness of the gecko? Probably not. Genetic research provides a clue as to why sex is more rigid in our species—why we couldn't handle the "Honey, it's a warm day; let's have a girl" approach. Larry Shapiro, a specialist in human genetics at the University of California, San Francisco, points out that such a system would be evolutionary whimsy—or perhaps suicide—for humans. We don't pump out hundreds of well-fertilized eggs a year. Barring twins—and the occasional rise to sextuplets—a human mother can produce one baby a year, tops. If we want the ideal mating ratio—a male for every female—then we need tighter control.

"When animals produce a large number of offspring, they can afford to leave sex determination to external factors," says Shapiro. "Not so, mammals and (especially) primates. Primates produce such a relatively small number of offspring that there has to be more control. Otherwise the ratio between males and females would become skewed." The worldwide sex ratio is fixed at approximately 50-50 for humans. Males tend to outnumber females slightly at birth; females tend to survive at a higher rate. It evens out. The numbers stay balanced.

At a basic level, this system, rooted in time, means males and females can never be biological mirrors. We can debate all kinds of fine variations in behavior; we can find enormous overlap between male and female behavior. But, barring genetic disorders or hormonal malfunctions, we get a clearly male or female body and a strong sense of gender identity. Our genes do not offer the option of sliding between male and female according to the environment. At some stage in evolutionary time, our species opted for a certain rigidity. Not that it wouldn't be interesting to see one's husband, say, switch to female because he could survive better at the office that way. Or vice versa. And not that we're cast in stone either. There's enough slip in our biology to allow for transsexualism, for variations in sexual orientation, and even leeway in the way our bodies get built. But not enough to make us identical counterparts.

Kim Wallen, a professor of psychology at the Yerkes Regional Primate Research Center, based at Emory University in Atlanta, puts it like this: "The effect of a predisposition is essentially to open up the path of least resistance. [For example,] our visual system is biased to develop in a certain way, based partly on certain environmental inputs." Many mammals, for instance, have a vision system that matures after birth, doing so normally only if the eyes open onto light. "We know from experiments with cats [raised in the dark as kittens] that we can alter the system," Wallen says, "but it takes special energy to do so, special environments, special conditions.

"Let's apply that to gender predispositions," he continues. "Let's take an extreme case—you want to make a male who can give birth to a baby. Well, you could do it. But it would mean all kinds of surgeries, developing and sustaining an artificial womb, years and years of development. In other words, a massive input of energy. The path of least resistance in our species is that males don't give birth and females do. A lot of our behaviors follow from that—and to change them, you have to be very determined."

It's the issue behind Wallen's statement that leads me to the subject of genes. I'm interested in how they shape us less as a biology lesson and more as a concept. How might genes make a man different from a woman? What might that imply for a range of behaviors?

The understanding of DNA—the discovery of its now-famous double-helix structure, the knowledge of its chemical composition—has revolutionized science. The massive Human Genome Project is built on the ability to pick apart that structure bit by bit, enumerating each crystalline base—adenine, cytosine, guanine, thymine—linked together, pair by pair, into a code. The bases are assigned a single-letter identity, corresponding to their names: A, C, G, and T. On paper, this actually starts to look like a coded message; in a scientific journal, the relaying of this code, for even a single gene, resembles the product of a four-letter typewriter with a stutter: CCCTAG/AAATTC/ and so on, with thousands of repeats spread over page after page.

For a geneticist, that's the world within the world, an amazing and genuinely beautiful tapestry. Each cell chromosome, then, is like an ingeniously assembled collection of these tapestries. The ability to decipher the code means that biologists can copy genes, rearrange them, search out ways to correct genetic defects, and work on the details of the connection between genes and behavior.

On the other hand, the four-letter genetic alphabet reads like gibberish to most of us. There's nothing wrong with that either; most of us won't need to spend our days ticking off base pairs in preparation for cloning a gene. What's far more crucial is for us to get comfortable with the ideas behind modern genetics—the growing understanding of where genes have great power, and where they have unexpectedly little.

Knowledge, of course, is a moving target. What we know about genes today will shift tomorrow. But I think we are now at a beautifully balanced point in the scientific concept. We are no longer so naive as to embrace complete genetic determinism, nor do we embrace its opposite. Biologists are deftly showing us not only that genes have incredible influence, but that they are, that we are, also designed to be highly responsive to life itself—so that life in turn can influence whatever genetic tapestry we bring into the world.

Once you get the concept down, you're protected against the genetic mind-games that some people like to play. You won't fall for the rather horrifying idea that a single gene rules your life, making you incurably stupid or fat, dooming you to cancer, or locking you into a sexual orientation. Genes don't work that way; biology doesn't work that way.

On the other hand, there's no doubt that genes define the species—you can't tweak the human genome into the rice genome. The genetic code spells out some basic laws of bodybuilding and body recognition. We don't understand them fully. How does a male learn to recognize a female, say, especially within a species, such as Canadian geese, where there's very little difference in outward appearance. It's one of the more interesting—and difficult—questions in biology.

In our species, genes code for proper and specific assemblage of human bone and muscle, skin and nerve. We know, though, that even at this most fundamental level, there's room for flexibility and for error, even lethal error. The estimated miscarriage rate for one-week-old human embryos is about 50 percent, most due to severe genetic errors in the developing body. University of California, Davis, anatomist Allen Enders once suggested that, based on such numbers, we should consider God the world's greatest abortionist.

Genes are a starting point. They're the manual, the chemical instructions for assembling amino acids. An assembly of amino acids is a protein. Those proteins, in turn, form membranes or hormones or muscles, or whatever is needed. To stay with the instruction manual analogy, think of genes as the sentences and chromosomes as the pages. The genome itself is the whole book, very precisely written and unforgiving of errors. Skip a sentence, mutate a gene, and the end result can be disaster. Humans have 46 chromosomes—23 matched pairs, an equal half from each parent—containing somewhere (we really have no idea) between 70,000 and 150,000 genes. The chromosome pairs are numbered according to size: one is the largest and 22 is a stub. The final pair holds the so-called X and Y chromosomes, which determine whether we are male or female.

Here's where the mathematical odds come in when working out the male-female ratios in humans. In mammals, females have two X chromosomes. Males have an X and a Y. When a sperm fertilizes an egg, chromosomes of both parents merge, each contributing half the genetic material of the child. That means that the father essentially determines the child's sex. The mother donates an X chromosome; the father provides either X or a Y, and if there's a system to determining which one it is, we haven't figured it out yet. So, it's a flip of the coin—roughly 50-50 odds on X or Y. The same system applies in birds, only in reverse. The mother determines the child's sex: males carry two Z chromosomes; females, a Z and a W. There's no obvious reason for assigning those letters to sex chromosomes; somebody liked the tail end of the alphabet, apparently. It's the concept that matters to us: the parent who carries the mixed package of sex chromosomes determines the sex of the child.

Genes start our lives; they give us a beginning. But neither they, nor we, are totally predetermined. Some genetic instructions, like eye color, are written in ink. But others that scientists once thought were totally preset have proved less rigid. Height, for instance, is proposed by genes but can be altered. Geneticists estimate that height is about 90 percent heritable, meaning that the environment can slightly redirect your body's genetic plans for height. Environment in this case is nutrition. Starve a baby whose genes meant him to grow to six feet tall and his adult self may fall short by an inch or more. His body cuts back, conserves, and stunts growth in favor of survival. There's even a new, somewhat controversial, study that says stress in childhood interferes with height, suppressing growth hormones. It suggests girls from troubled homes grow up to be smaller women.

Genes themselves, it appears, are programmed to react and interact with environment; nutrition and weather, internal chemistry and external, all can affect our genes. Our bodies are acutely tuned to the world. The example of birdsong that Emily Dickinson so wryly offered is actually a wonderful demonstration. The lark, the sparrow, the yellow canary—all are genetically set to sing, in contrast to the busy cluck of a chicken or the war cry of a red-tailed hawk. Baby songbirds know the melodies without being coached. And yet the genetic background can depend on environmental input. The young bird must hear the song of its own species at a very precise time. If it hears the wrong song—a lark accidentally eavesdropping on sparrows—it garbles the tune. In a sense that dooms the bird, because it has lost its chance at learning the notes to seduce a mate. If white-crowned sparrows don't hear their song at the right time, they never sing at all. It's comparable to Wallen's point about developing vision—if kittens don't see light at the right time, they may not see at all.

Just as birds and kittens are designed to interact with life, so are we. We know there are "windows of opportunity" for development. The best studied is language. If children don't hear words, or use them, before age five, a full grasp of language seems to elude them. The part of their brain that processes words doesn't develop properly. When words came slowly to one of my nephews, doctors told his mother, my sister, to learn and use sign language. They wanted to coax his brain, somehow resistant to verbal cues, to engage in language before he was beyond his crucial receptivity to it. It was important not to wait. The tactic worked. Now he talks with the rapid-patter exuberance of any five-year-old.

The story of songbirds, though, raises yet another aspect of genetics. While the idea of an opportunity window allows for genetic response to environment, it's not necessarily an obvious response.

Genetic responses are not direct, especially in the way they influence behavior, as evolutionary biologist Judy Stamps points out. It's not as if you have a gene that turns on a song, for instance. What genes do, really, is produce the machinery that allows a bird to sing. And that's a complicated process: a tiny male bird hears a song, which initiates gene expression, which, in the presence of testosterone, starts a production line to organize the neurons to make a special song center. Ever after, the bird can sing himself (there's evidence that they practice their notes) and trill out the song of his species.

Obviously, if the bird never hears the song, this doesn't happen. Or, if you block testosterone in a baby zebra finch, it doesn't matter if he hears a whole operatic suite from his fellow finches—he doesn't sing. The genes may have correctly assembled the testosterone response, but if a scientist stops the hormone from reaching the brain, the gene's work is undone, and the signal to create the bird's own song abilities is never heard.

Stamps offers another illustration of genes and behavior interacting, from the insect world: "At the risk of vast oversimplification, a female katydid's genes might encourage the development of neuronal (nervous system circuits) and hormonal systems. These would allow her to monitor food intake levels, egg production rates, encounter rates with other females, and encounter rates with reproductive males. She could use this information to adjust the amount of aggression she directs toward other females or the sexual behavior she directs toward reproductive males." So genes indirectly create a system that allows the katydid to behave differently every day.

If we narrow the emphasis to gender differences, there are some very preliminary suggestions that genes may act differently on a male versus a female chromosome—that the female chromosomes may be a little more active somehow. This is so new, however, that no one knows exactly what it means. To try to figure that out, biologists are looking at what genes do—the end products that they code for. In exploring sex, they've been drawn particularly to hormones. Hormones are chemical messengers; they signal each other—they signal changes in cells ranging from nerve to bone. Sometimes one hormone alone seems to influence a behavior; sometimes it appears to be a vast hormonal army. They are, as a group, as busy a piece of biology as can be found. Hormones are released in the brain, by glands, by fat tissues, by the ovaries, by the testes. They do an incredible amount of work—adjusting blood pressure, settling us into sleep, helping build muscle structure, strengthening bones, driving sexual desire. We focus so much on the behavioral aspects of hormones that we sometimes forget how much they also keep us alive.

In behavior, at least in the context of sex differences, the critical hormones are steroid hormones, which are actually derivatives of cholesterol. The so-called sex hormones are called androgens and estrogens. Androgens, from the Greek *andros,* for male, circulate at the highest levels in males. They are made predominantly in the testes, providing the name for the best known of them, testosterone. Estrogens, from the Greek *oistros,* for gadfly (is this sexist or what?), circulate at the highest levels in females. They are made mainly in the ovaries. There are three main estrogens; the one we know best is called estradiol, and it dominates until menopause, except during pregnancy.

None of the hormones are gender exclusive. Males make estradiol and females make testosterone, mostly in the adrenal gland. The brain can convert testosterone into estradiol as well. The difference is in how much hormone circulates in the bloodstream. Men average about ten times as much testosterone as women. Estradiol occurs in exactly the opposite distribution. The question, and I do mean THE question, is whether the hormones organize human brains as straightforwardly as they appear to do in other species. It brings us to yet another take on birdsong: Males are the musicians of the songbirds. But at any time during the life of a female canary, she can be induced to sing simply by injecting a male-level dose of testosterone.

No, it isn't simple. This is a complicated formula. The particular genes you inherit, your fluctuating hormone levels, your individual environment—your ever-changing life—all work in concert and separately. Further, genes can also simply make mistakes, and randomly mutate. The typewriter key sticks, you get an extra C, and suddenly the system scrambles. What if you're fixed on the biology of, say, a killer. Can you argue that genes, those assemblers of amino acids, made him deadly? What if they assembled an extra-aggressive testosterone? On the other hand, testosterone is extraordinarily sensitive to environment, rising in times of challenge and plummeting as an argument is lost, climbing again with sexual desire. If witnesses describe this hypothetical murderer as unusually jittery, was it the genes, the hormones, a really lousy day, or a burst of lust? This isn't just a male question; there's some evidence that testosterone fluctuates in women too, depending on how stressful their lives are. You can see why some geneticists find code translation—scrolling through the CCAATs—a lot easier.

On the other hand, testosterone performs some straightforward and unmysterious work. There's a solid consensus among scientists that, in humans, this is the hormone which, at least at the beginning, separates the boys from the girls.

In the human embryo, if there's no introduction of testosterone, you get a girl. As with birdsong, the genes signal hormones to do the specific work. So—as we'll discuss in detail later—you can have the classic male XY chromosome setup and still produce a person who looks female if testosterone is blocked. Thus, some scientists describe the basic mammalian structure as female—the idea being that all of us would become females if testosterone were not plugged in.

There are a number of very well-known sex researchers who call females the "default" sex. Marc Breedlove, at the University of California at Berkeley, hates this idea. To him, that suggests that, to become a boy, you have to "improve upon" the female structure. "As if boys are worth some extra bells and whistles," he snaps.

Curiously, the system works the other way in birds. If you can live with the idea of a default sex, then in birds, the default sex is male. How do we know this? Male and female birds are often gaudily different. Think of peacocks, pheasants, and ducks: the iridescent fan of a male peacock's tail, the bronzy gold plumage of a male pheasant, the glimmering green head of a male mallard—and the dusty, dowdy browns of the females. The colors have all kinds of meanings. They tell us, for instance, that these particular males don't share in incubation duties; their very brilliance would call predators to the nest. But more to the point, the female's coloring is set by genes that turn on estrogen in the she-chicks. Suppress the estrogen—remove the ovaries of a female peacock—and she'll blossom into male splendor. Sometimes it happens anyway; female peahens with ovarian disease will sprout flashy plumage. In other words, male is the basic approach for birds, just as female appears to be for mammals. The clear connection is to the sex with the matched sex chromosomes—XX in mammals for female, ZZ in birds for male.

The double X is so formidable, in fact, and the Y chromosome so insignificant, that there is some suggestion that, given enough time, we'll all end up as variations on an X-related theme. Scientific journals tend, these days, to publish articles with titles such as "The Rise and Fall of the Y Chromosome," or "Whither the Y," or even "Equality for X Chromosomes." The point is raised by evolutionary biologist Jim Bull, who asked, almost irritably, in a 1994 *Science* magazine article, "Why the hell does the Y chromosome end up as a little blob with hardly any genetic activity on it?" X chromosomes are loaded with the right stuff. Each X has somewhere between 2,500 and 5,000 genes, including those that code for muscle development, blood clotting, and color vision. The Y, at this point, seems to hold only about 15 genes. The main purpose of the Y is that it holds a Sex Determining Region, as some call it, or the TDF—the Testis Determining Factor. The gene or genes in this region are responsible for the initial "go male" signal. That causes the developing human to build the testes, which in turn make testosterone and other androgens, which in turn act as messengers to build a male body.

On this point, you can't underestimate the importance of either testosterone or the Y chromosome. And yet the Y chromosome, because it's become so focused on one task, puts males at a disadvantage. To use the computer analogy, we're all better off with backup. Say you inherit an X chromosome with a defective gene. If you also inherit another X with a solidly healthy version of the same gene, it's the good copy that slots into place. But if you are XY, there is no alternate. The male is stuck, which is exactly why men suffer more from X-linked disorders—color blindness, hemophilia, Duchenne muscular dystrophy, and fragile-X retardation. There are more than 200 of these disorders, and they all occur more often in men than women. If a female develops one of them, it's because she had the bad luck to inherit the same malfunctioning gene on both X chromosomes. It's partly here, in this genetic inequality, that the pattern of early male death begins, and it's that pattern that Don Grayson thinks is magnified by disasters like that of the Donner Party.

Most scientists suspect that the two sex chromosomes used to be equally matched. But as their roles became more entrenched—you go male, I go female—eventually they stopped consistently swapping genes. This effectively isolated the Y chromosome; the X, however, still continued a healthy exchange when it matched up with another X, in the female mode. As the Y had less and less to share, it became more and more isolated, cut off from fresh genetic material. Eventually it began to wither down to a mostly single-purpose tool.

Biologists, using fruit flies, have shown how perilous this is for a chromosome. In some 35 or so fruit fly generations, an isolated chromosome will evolve out of existence. Does this mean we're evolving toward asexuality? Not really. Fruit flies, in fact, simply use X chromosomes to determine sex: XX for female, single X for male. They preserve, though, that unwavering dedication to the system of two sexes.

We've known for about 30 years that a small subset of human males are born with an XX chromosome set. Back in 1966, this was a major shock. It went against all logic and all understanding of genetics—against what was perceived as the ironclad division between males and females. Of all the self/nonself rules going, the he-she one seemed unbreakable. As it turned out, researchers found that the XX males had, by a rare genetic misstep, a tiny slice of the Y chromosome attached to one of their Xs, representing about half of one percent of the genes that normally lie on the Y. By deduction, they realized that this must be the Sex-Determining Region of the Y chromosome.

By experimenting with mice, British scientists have shown that if this particular slice of Y is removed from a classic XY male, the animal does become female. They also showed that in normal male mice, this region lights up and gets active just at the point of sexual divergence in the embryo. On the other hand, further research has also found that the XX males are missing some critical piece of the information. Those particular mice don't get fully developed male genitals; neither do XX men. There's obviously more to the Y chromosome than we yet understand.

Although such studies tell us that we share some genetic principles with mice, obviously we develop along a very different program. Here's the basic schedule of early human life: For the first six weeks of development, human embryos grow in a basically gender-neutral route, a blob of cells with unlimited potential. During that time, the embryo makes gonads of no particular sex, just a foundation for what will come. If the embryo has a Y chromosome (or, as suggested earlier, at least a key part of one), then the relevant genes send out a signal for testosterone. That sort of rising flare cues the cells into a frenzy of androgen-aided building. A male embryo, at six weeks, starts converting those gonadal lumps into testes which make testosterone.

Female embryos don't start developing ovaries until about 12 weeks. But even before that, a female future is settling into place. It is during this shadowy time frame, when the emerging embryo is still tiny and unformed, that all the egg cells she will ever have are made—300 to 400 that may be released in her lifetime, perhaps

100,000 that will not—and tucked carefully away, waiting for someday's mating season.

If you examine an embryo at about six weeks, you see that it actually has the ability to develop in either direction, another reminder that we are different sexes but that we all exist along the spectrum of human possibility. The fledgling embryo has two sets of ducts—Wolffian for male, Muellerian for female—an either/or infrastructure, held in readiness. At this stage, hormones seem to function something like a construction supervisor. If the embryo goes male, the internal ductwork for a male enlarges and the female structures begin to shrink away. Testosterone causes the Wolffian ducts to enlarge; these develop into the vasa deferentia, epididymis, and seminal vesicles. In simpler words, they become the delivery channel leading from the testes to the penis, carrying sperm where it needs to go.

At the same time, the testes put out a kind of antifemale hormone called Muellerian regression, or Muellerian inhibiting factor (MIF). This causes the interior female ductwork, still on standby in the male, to finally shrink away. There is some suggestion that the Y chromosome may also code for this uniquely lethal formula. The ovaries do not make a Wolffian removal formula; in fact, they produce no duct-killing chemistry whatsoever. They don't have to be nearly so aggressive about removing traces of the opposite sex—without the androgens, the Wolffian ducts die from simple lack of support. In females, the Muellerian ducts continue on to become oviducts, uterus, and vagina—the comparable delivery system and the network that will, one day, form the pathway for an egg to be fertilized, nurtured, and delivered, all designed to begin the cycle again.

Hormones help wrap the package as well, adding the "bells and whistles." It's an extraordinarily ordered system, as one can see by looking at the precision building of a penis and scrotum. Here, apparently, simple testosterone power isn't enough. The genital skin contains a hot little enzyme, 5-alpha-reductase, that converts testosterone into dihydrotestosterone. This revved-up androgen binds even more tightly to the cells, producing a focused, pinpoint-concentration effort. The body spends a little extra effort trying to make sure that the penis is functional.

Marc Breedlove has demonstrated in remarkable detail how this kind of sculpting occurs. Breedlove's study focused on a structure called the spinal nucleus of the bulbocavernosus (SNB), in the spinal cord. This is a cluster of neurons, or nerve cells, that control the bulbocavernosus muscles at the base of the penis. Breedlove experimented with rats (imagine trying to persuade a university committee to let you do this in humans; even better, imagine trying to line up volunteers). In rodents, he found that both males and females look alike as embryos: they have an identical muscle and a matched set of neurons to control it. By birth, though, only the males have the muscle. They need it; a rat's penis has to have enough punch to break through cervical plugs in the female. By contrast, before the young females are born, both the neurons and the muscle disappear entirely. The difference, Breedlove discovered, was strictly the result of testosterone at work. The hormone provided a signal that stimulated muscle growth. The muscle, in turn, sent its own cue to the waiting neurons, alerting them that they were needed. Without that alert, the neurons are programmed to self-destruct.

There's a direct parallel to humans. We also have both an SNB and a much-valued bulbocavernosus muscle. In men, the muscle wraps around the base of the penis, as it does in rats, and aids in semen ejaculation. In women, it encircles the opening of the vagina and helps constrict it. So, unlike rats, we all keep it. But in humans, the muscle is much larger in men. So is the nucleus in the spinal cord. Men load up some 25 percent more neurons there than women do. And as in rodents, this sexual split occurs before birth. About the 26th week of human gestation, if testosterone is circulating, the male muscle and nerve system gets built. Without it, the female's becomes compacted. The consequences, if this happened in reverse, are pretty easy, and somewhat painful, to imagine.

The building of sexual structures such as the penis is pretty unambiguously linked to hormones. But after birth and throughout most of childhood, the steroid hormones fade to the background. It's another reminder that their primary purpose is sex—reproduction if you like—before all. They spike again in puberty, when their handiwork is very visible—and very familiar. With rising levels of androgens, boys bulk up; their voices change; they grow swathes of hair. Girls grow fully rounded breasts, a distinctly human characteristic among the primates. Girls also acquire the curved shape associated with a new distribution of fat—that protective layer mentioned previously—and they begin menstruating. There is a blueprint here. It's the same one charted by Breedlove's work, and has to do with functional reproduction.

For all the research, though, we really don't understand fully the reasons behind the design. Scientists have never explained the unique shape of a woman's breast. This shape is not needed for milk production, as a comparison with any other primate shows. At best, they've come up with the somewhat obvious idea that men find big breasts seductive, encouraging their formation during evolution. You can tell, though, how little anyone likes this theory by the fact that hopeful alternatives keep appearing in the research journals.

And dwelling on the penis for a little longer, so to speak—one of the evolutionary puzzles has been the length of the human penis compared to that of other primates. UCLA physiologist Jared Diamond writes, jokingly, of the great failure of modern science to come up with an "Adequate Theory of Penis Length." Gorillas grow thimble-sized penises—only 1.25 inches on an average erection—and still successfully manage to get a female gorilla pregnant. So why, exactly, does a human male need five inches or greater? Diamond favors the idea that somewhere in our past this was, in fact, a threat display. You know, deer have their antlers, sharks have their teeth, and men have their— Well, you get the idea. In other words, scientists, like poets, still find sex a mysterious entity.

But clearly, if a body doesn't get the sex-appropriate hormonal setup, it doesn't get the appropriate structure either. For some 50 years now, scientists have been growing female structures in males, and vice versa, by playing with hormones. In 1947, the famed biologist Alfred Jost castrated male rabbit fetuses—thus cutting off their testosterone supply—and watched as they produced vaginas instead of penises. It's the same approach with birds: block the estradiol and suddenly the dowdy female blazes into male glory.

We don't do these experiments in humans—for obvious reasons—but there are naturally occurring instances that prove the same point. For instance: the fetal testes are composed basically of two kinds of cells, called Sertoli cells and Leydig cells. Sertoli cells are the ladykillers; they make the anti-Muellerian factor that dispenses with all the unwanted female plumbing. The Leydig cells make testosterone. Their job in a male fetus is to pump the hormone out. In rare cases, the fetus simply doesn't produce enough Leydigs (a defect known as Leydig Cell Hypoplasia). This defect begins in the pituitary gland, part of the testosterone relay system. The pituitary fails to send a strong signal; Leydig cells don't form properly, and the testosterone ball gets dropped. Although the fetus is a fully qualified XY boy, he never grows a penis.

Obstetrician and parents think they see a baby girl. The child is raised as a daughter. The mistake gets caught about the time of puberty, when menstruation doesn't start. A doctor's examination shows the poor kid to be internally male, without uterus or fallopian tubes (the Sertoli cells quite properly dispensed with those). If the doctor looks further, there are usually small testes, often tucked within the abdomen. As the researchers put it, if the condition had been known from the beginning, "the sisters would have been born as brothers."

There's another defect in which testosterone pumps out all right, but the body can't detect it; the receptors are missing. Think of this as a kind of hormonal deafness: the hormone is talking but the receiver is turned off. Called Testicular Feminization Mutation, it's caused by a defective gene for androgen receptors. In this case, although the baby once again has the male XY, it's born with a clitoris, labia, and shallow vagina.

To complicate that even further, these children do develop breasts at puberty, in a way that emphasizes how absolutely precise the hormonal balance must be. It appears that human breast development is based on a kind of internal math relating androgens to estrogens. The higher the proportion of estrogens, the more the breast swells outward. Now, these particular children have the normal male amount of estradiol, about a tenth of that found in females. But against the lost testosterone signal, the body reads the estrogen cue as female.

Some girls are born with a malfunction of the adrenal gland, which in females is the primary producer of androgens. These girls lack an enzyme which normally converts most of the androgens into cortisol, a hormone that regulates metabolism and stress. Girls with this defect—called Congenital Adrenal Hyperplasia (CAH)—are awash in androgens laid on top of the standard female chemistry. As a result, their bodies receive an incredibly mixed signal. And it shows. CAH girls are readily identified at birth, with a large, penislike clitoris and a partial scrotum. They are treated with surgery and medications to adjust the hormone balance.

Finally, just to emphasize how tricky all this bodybuilding can get, there's a peculiar genetic defect that seems to be clustered by heredity in a small group of villages in the Dominican Republic. The result of the defect is a failure to produce the enzyme 5-alpha-reductase. Remember that this enzyme is responsible for the intense local signaling that builds penis and scrotum, a built-in amplifier of the testosterone signal. Here we see how essential the enzyme is. Without it, you get a boy with undescended testes and a penis so short and stubby that it resembles an oversized clitoris. The children are usually raised as conditional girls. At puberty, the secondary tide of androgens rises and is apparently enough to finish the building job. The scrotum suddenly descends, the phallus grows, and the child develops a distinctly male body—narrow hips, muscular build, and eventually even slight beard growth.

At that point, the family shifts the child over from daughter to son. He wears male clothes and starts dating. People in the Dominican Republic are so familiar with this condition that there's a colloquial name for it: The children are called guevedoces, meaning "eggs (or testes) at 12." I suspect, because of both the frequency of the phenomenon and its self-correction, that their transition is less traumatic than that of those "sisters who would have been born brothers." Still, identity for most of us is grounded in our gender. It must be unnerving in the extreme to have that slide out from under one's feet.

These are all rare conditions. They testify to the influence of hormones in making bodies. They don't alter the basic male-female division of our species, but they emphasize how fragile those divisions can be. There is no supermale security in possessing a Y chromosome, no femininity guarantee with the double X. Biology allows itself mistakes. It allows us flexibility. With that gift comes the risk of bending too far. That it's possible to live with these genetic defects, that they don't merely kill us off, is a reminder that we are one and the same species. Male and female alike, we exist on a continuum of biological possibilities that can overlap and can sustain either sex. How do we ultimately define a male or female —by body parts or behavior? Or do they wrap as tightly together as the double strands of DNA?

The most perplexing body part of all, of course, is the one within the skull. The obvious next question is whether the nervous system also differentiates at this level—whether we get distinctly male and female brains to accompany these distinctly male and female body parts. At one level, this may seem a ridiculous question. There's no reason to think hormones neutralize above the neck. There's every reason to suspect that you might need different controls in the brain for the different parts. Breedlove's work on the spinal cord shows clearly that hormones can sculpt the central nervous system. However, it also shows that you don't necessarily need to involve the brain with every body function or every flex of a muscle.

Breedlove himself points out that the most difficult task may be differentiating how the brain responds directly to hormones from how the brain responds to the *results* of hormones. The penis, in this context, can be considered a result. Before birth, the androgens, secreted by the testes, help build a penis in a boy. "And after birth," says Breedlove, "virtually everyone who interacts with that individual will note that he has a penis and will, in many instances, behave differently than if the individual were a female. These experiences will to some extent shape brain development and result in different behaviors from childhood to adulthood.

"I find it impossible to classify this chain of events as either purely social or purely biological," Breedlove adds. "Even the most ardent proponent of inborn sex differences in behavior would admit that ignoring the penis and raising the individual as a girl would have some effect on development."

Perhaps at this point, we'd do best to fall back on the poet-playwright: As Shakespeare's Hamlet exclaimed, "What a piece of work is man."

Chapter Two

PINPOINTING THE DIFFERENCE

Comparing Male and Female Brains

Imagine yourself in an autopsy room, awash with fluorescent light and the stinging smell of disinfectant. Two bodies, one male and one female, are laid out for dissection on metal tables. It may take a pathologist's degree to determine cause of death—did the heart falter first? If so, in which valve?—but, given a body in reasonable shape, none of us needs medical school to determine sex. It's so obvious that it hardly requires thought; we all recognize the generalities of shape and size and parts. We could do it by piece—a leg, an arm. (I know it's gruesome, but stay with me.) Most of us could pick up a foot, at least an adult foot, and identify it by sex.

That's not true for every body part. Even medical school won't teach most students to sex a heart, a liver, or a kidney on sight. Why should it? We are one species; there's no obvious reason why a man's heart should pump differently from a woman's. Transplant surgeons know gender is not the issue in organ recycling. If a heart is going to be transplanted, let it be these things: a good immunological match, a shared blood type, and healthy; let it flex, beat, and thump the blood out. When it's stitched into place in that waiting hollow, let it save a life. Male or female doesn't matter. Foremost, it's a human heart.

In our autopsy room, of course, it's too late to talk of saving lives. The hearts are silent in those silent bodies. We're looking for something else. We carefully open up each skull, peel away the thin gray membrane of the dura, detach each brain, and set the two of them side by side.

There is no "Aha." No "Look, a male brain, armed for battle," or "A female brain, made for motherhood." While it's impossible to look at a brain—its faint, reflective gleam, the intricate coiling structure—and not marvel, you can, as with the heart, stare at those convolutions from dawn to dusk, even dissect them, without seeing sex. A human brain is not, like an arm or a leg, obviously male or female.

Why not? Scientists know from looking at other species that if the sexes behave very differently, their brains become visibly different. The classic example is songbirds. The sexual difference here is song itself. Males are the musicians, mostly. (There are some duetting exceptions.) They use their melody sometimes as warning but mostly to court—to advertise themselves to available females. That silver song that Emily Dickinson so admired—"Saved for your Ear, when Lutes be Old"—that trill of melody that wakes me up before the sun, before I am ready to be a reasonable human being, is evidence that male birds have sex on their minds (and far too early).

Because male songbirds do sing—and their mates do not—the males use their brains differently to produce music. The "music box" of their brain sits near the front. It's usually called the Higher Vocal Center. In this compact nucleus, the neurons of males usually expand into a space some six times greater than that of females. The difference is so strong that scientists who study songbirds can see it when they do a simple dissection of the brain.

Nothing in humans compares. There is an overall size difference: by weighing and measuring hundreds of human brains, researchers have found that, in general, men's brains are about 15 percent larger than women's brains. This means an average difference of about three ounces, in brains that weigh around 2.5 pounds. The logical assumption is that men need a bigger brain to move those bigger bodies. But, in fact, the parts of the brain that control large muscle movement are not outsized in men and do not seem to account for the overall difference.

More than any other gender comparison in biology, it's fair to say that feminist scholars hate this one the most. Brown University geneticist Anne Fausto-Sterling argues that the work is biased from its start; male scientists consistently find that male scientists have the biggest brains. Since we tend to assume that bigger is better, the implications are obvious.

A hundred years ago, human brain-size comparisons were the cornerstone of a scientific argument that women were hopelessly more stupid than men. But, then, every visible difference used to be written up as evidence of women's inferiority. Even the more delicate faces of women were translated as childlike and "naive." At best, said British biologist Havelock Ellis, a woman's finely drawn features might suggest "a face turned upwards to kiss." In that context, nineteenth century scientists, almost exclusively male, took the brain-size difference as simple proof of the masculine advantage. Some explained that women had failed to evolve as far as men. They suggested grouping women with the lower primates—gorillas coming to mind—in terms of intelligence. The great nineteenth century neuroscientist Paul Broca was more moderate, but even he put it this way: "We might ask if the small size of the female brain depends exclusively on the small size of her body. . . . But we must not forget that women are, on the average, a little less intelligent than men, a difference that we should not exaggerate, but which is nonetheless real."

In our own century, at least in the last twenty years, most scientists have expressed puzzlement. Some suggest that women don't need larger brains, that we've developed instead a system more densely packed with neurons and more efficient. There are feminist scholars who simply say this preoccupation with size is meaningless, that the best policy is to declare the difference irrelevant and leave the topic alone. This seems to me a risky approach, particularly for us women. The difference is there; better to explore it, figure it out, and explain it rationally than to wish it away. Failure to do so leaves a gap that certain scientists—a vocal, almost exclusively male minority that is happy to keep women "in their place"—will cheerfully fill.

In fact, they already have. Not surprisingly, these scientists turn out to be the same researchers who argue that whites are naturally smarter than blacks. J. Philippe Rushton, of Canada's University of Western Ontario, has dedicated entire books to the notion of black inferiority. He also argued in a 1996 paper, "Brain Size and Cognitive Ability," that the conclusions of Broca "and other nineteenth century visionaries" were right on—that differences in brain size are central to intelligence. In part, his argument follows the idea—which we'll explore further in this chapter—that men are better at spatial reasoning. A close colleague of Rushton's, psychologist Richard Lynn of the University of Belfast in Northern Ireland, makes a comparison with computers: Spatial analysis and graphic processing require a lot of space.

It's not a ridiculous analogy if you accept that, overall, men are better than women at handling numbers and maps and spatial analysis. But Lynn likes to push the point as evidence of a broader gap in real ability: "If I think of the very intelligent people that I know, more of them are men," he says, suggesting that if women just had slightly larger brains, they'd score better on IQ tests.

He, Rushton, and colleagues seem belligerently out of their era. Their work flows from the nineteenth century devotion to proving the white male supreme, especially the northern European white male. In doing so, they let the perspective of today's science pass them by. In the nineteenth century, microscopes weren't as powerful as they are today. Now we can move from the autopsy room to a laboratory where it's possible to magnify a human nerve cell to 100 million times its actual size. We find even at that level that the brains of men and women are almost but not quite mirror images—there are tiny, pinpoint differences within. And the pinpoints are far more intriguing.

Pick the brain apart at a cellular level (some studies read as if an accountant had been set loose in the neural circuits) and you can find a distinct set of gender differences. Scientists call them sexual dimorphisms—literally, two shapes for two sexes. The differences are more subtle in us than in songbirds; there's no ostentatious song-control center. But they're real. They're also very difficult to pin down, because—contrary to the beliefs of those nineteenth century champions of maleness—the brain is not a constant. It spends much of its life improving itself, refining the structures and responding to internal and external changes. And male and female brains appear to be slightly different in their response to changes. For example, men's brains, though larger overall to begin with, shrink faster with age. This seems to make Lynn's quest to prove that a larger brain equals a smarter one literally a pursuit of a moving target.

Some very new studies, mostly in rodents, find that during ovulation the female's brain is incredibly responsive to the rising level of estradiol. Within the hippocampus—a structure that handles memory and which, incidentally, is linked to spatial processing—the neurons send out a sudden rush of new branches, called dendrites. As the ovulation cycle closes down and the hormone levels drop, the dendrites retract. If a scientist were scrutinizing that brain, it might look startlingly different at the dendritic level, depending on the time of the month. Some research on how responsive the brain is, most notably by Marian Diamond at the University of California, Berkeley, has shown that even in elderly male *or* female animals—rats with an age comparable to a 90-year-old human—you can produce new nerve growth in the brain simply by introducing new challenges into the environment.

"People are finding that the mature brain is extraordinarily plastic," says David Amaral, a neuroanatomist at the University of California, Davis. "Dendrites are growing and retracting, synapses are falling apart—there's a lot more potential for growth and change than we used to realize."

Even the flashy sex differences in songbirds aren't stable. The song nucleus expands in the spring—the musical, lyrical breeding season—and contracts in the fall. If you dose female canaries with androgens, they will trill away. It will also cause them to grow a song center. But that nucleus tends to be at best about half the size of the male's. Researchers thus puzzle over why the female canaries can perform like males with only a fraction of the specialized neurons. No one knows, but again it adds an interesting twist to the argument that size equals ability equals male superiority.

In other words, nailing down a clear structural difference in the brain is fraught with complication, partly because the structures themselves are not always constant in this realm of the supersmall, and partly because scientists do not fully understand the relationship between function (the brain at work) and structure or morphology (the architecture within). It's even fair to ask what comes first: Does structure cause the brain to function in a certain way? Or does the way the brain functions cause it to build an appropriate structure?

Amaral, one of the most-respected neuroanatomists working today, puts it simply: "The linkage between function and morphology is pretty elusive." One of his favorite examples of the difficulty comes from an ongoing study by Michael Merzenich at the University of California at San Francisco involving mapping the somatosensory cortex of the brain. This thin surface region receives sensory signals from throughout the body that you can map: for example, the firing of neurons in the brain in response to pressure on a hand, a foot, or a finger. We know from these maps of electrical response that the fingers are rich in sensitive nerve endings, and that the back has very few.

Merzenich's truly brilliant work has also shown that if a person loses a finger, the firing pattern changes in the cortex: other fingers apparently become more sensitive to make up for the loss, as the brain's energy concentrates around the remaining digits. You can measure this phenomenon by comparing the response to the grip of a five-fingered hand to that of one with only four; the electrical crackle of nerve cells is adjusted. Yet the scientists have not measured any corresponding structural change in the brain itself. The sensory cortices, in terms of shape and size, look identical no matter how many fingers there are. From watching the brain at work, everything says that there's been a reorganization. But if it builds a new complex of cells, researchers haven't been able to see it. "And they've been looking for the differences for 15 years," Amaral says, "which tells you how hard they are to find."

When it comes to looking for gender differences, no part of the brain has received more attention than a curved cluster of nerve cells called the hypothalamus. The hypothalamus serves as a connecting station between the endocrine system, with its elaborate relay of glands and hormones, and the ever-responsive nervous system. The hypothalamus is a multipurpose structure if ever there was one. It sits in the front-center of all mammal brains, male and female, working like crazy to monitor internal conditions in the body. In its control are an eclectic, strangely connected, and absolutely essential set of life functions. They include hunger, thirst, body temperature, ovulation in females, and libido for both sexes. In the case of libido, the hypothalamus sends out a signal that apparently flags the steroid hormones into arousal mode. The fact that one structure juggles all these balls provides a neat biological explanation for why hunger and thirst interfere so directly with sexual interest. If this overworked little knot of cells is busy dealing with appetites more immediately essential to staying alive, it damps down arousal as a waste of time.

The hypothalamus also serves as a reminder that our biology often functions as a chain reaction, making it difficult to hang too much on one link. Ovulation, for instance, begins in this knotty little clump of cells, but not terribly obviously. The hypothalamus produces a hormone, called gonadotropin-releasing hormone, which nudges the neighboring pituitary gland into making another, called luteinizing hormone. That tells the ovaries to get busy and put an egg into production. There appears to be a comparable signaling system for men, in which luteinizing hormone orders testosterone production. Testosterone pretty consistently rises with the approach of a female, especially a sexually receptive one. So the hypothalamus is a logical player in the mix and match of mating systems.

It's logical, too, that when scientists began looking at minute structural differences in female and male brains, they began in this bossy little relay station. By the 1960s, the pendulum of scientific thought had swung to the opposite extreme from Paul Broca, and, except for the annoyingly persistent difference in overall size, the popular belief became that men's and women's brains mirrored each other.

But if mainstream science was never challenged, people would still be avoiding the night air for fear of inhaling malaria. So, in the late 1960s, two Oxford University anatomists, Geoffrey Raisman and Pauline Field, decided to test the mirror-image theory. They chose the hypothalamus as their starting place.

They focused on a tiny cell knot toward the front, called the preoptic area. The preoptic area (POA) is closely linked with male reproductive behaviors. If surgeons slice it out in rats, cats, dogs, goats, and monkeys, the male animals lose all interest in copulation. Raisman and Field chose rats for their experiment. They isolated the preoptic area and then they counted synapses, those lively points of connection between nerve cells. Actually, they looked at two different kinds of synapses within the preoptic: those that connect neurons and those that pass messages between branching extensions of nerve cells, such as axons and dendrites. The study is legendary, partly because it was such incredibly hard work, but also because it shocked the neuroscience community. At this level of fine detail, Raisman and Field found a gender difference: there were more type one synapses (the type that carry impulses cell-to-cell) within the POA in females. Suddenly, other researchers decided to look for subtle gender differences in the brain.

Since then, biologists have been picking away at the hypothalamus with, well, frustrating results. For example, in the front of the hypothalamus, there is clearly one cubic millimeter of neurons which are extremely sensitive to steroid hormones. This area is so different in males and females that scientists have named it the

Sexually Dimorphic Nucleus (SDN). Depending on which study you believe (all right, it's hard to count cells perfectly), the SDN is either five, six, or eight times larger in male rats than females—basically the difference between the head of the standard straight pin and the head of a fancy straight pin with a colored ball on the end. If you stain the cells with dye and know the territory, you can see the difference with the naked eye. Maybe there's a little squinting involved, but you don't need a microscope.

Work by neuroscientist Roger Gorski, of the University of California, Los Angeles, and his colleagues has shown that, in early development, androgens play a key role in shaping that difference. If you withhold testosterone from baby male rats, they end up with a small, female-looking SDN. Half the neurons promptly die. And if you boot up the androgens in females, they get a malelike SDN. What has been frustrating about the SDN is that so far the structure itself appears meaningless. Sure, it sits in the hypothalamus, in a region known for its sexual ramifications. Sure, it's responsive to sex hormones. But if you go into the brains of rats and carve out the SDN portion of the preoptic area, the animals pretty much behave like average rats; there's little or no change in the animals' sexual behavior.

Let's now complicate this further. Is there a comparable difference in human brains? One European lab, after autopsying a series of brains, reported that the SDN was about 2.5 times larger in men than women. That hasn't been confirmed, though. Gorski's lab, attempting an identical analysis, opened the hypothalamus where the SDN should sit and found no difference. What Gorski's laboratory found instead, primarily through the work of collaborator Laura Allen, was that the SDN was in all cases a distinct nucleus. And, if you looked around in this front region of the hypothalamus, there were three other obvious pencil-dot-sized nerve cell groups, each standing out against the background. Allen and Gorski named them the Interstitial Nuclei of the Anterior Hypothalamus (INAH) and numbered them in a series, one through four. They then decided to compare them in men and women.

They found that INAH-1 is comparable to the so-called Sexually Dimorphic Nucleus in rats, but the evidence for sexual dimorphism in that structure in humans is ambiguous. The Europeans—notably the laboratory of Dutch neuroscientist Dick Swaab—say it does exist. The Americans say no (although Gorski's study did look at older brains and found some evidence that this group of cells may equal out during a lifetime). Allen and Gorski found no male-female differences in INAH-4. But in the middle sets, INAH-2 and -3, there was something of a pattern: the nuclei were slightly larger in men. Again, no one knows what this means. These things aren't associated with any behavior—not clearly, anyway. But the reason we're bothering with this whole group of letters and numbers is that INAH-3, for those who follow sexual biology, is a fairly famous set of neurons.

It certainly brought one man to the forefront—Southern California neuroscientist Simon LeVay. In 1991, LeVay, already well known in scientific circles for his detailed and elegant studies of the nervous system, published a paper in *Science* magazine that basically reported on cell-counting across the INAH spectrum. LeVay didn't just compare men and women; he compared the brains of homosexual men with those of heterosexual men. He found no difference in INAH-1, -2, or -4. But in INAH-3, homosexual men tended to have fewer neurons—more than women, but less than straight men. This was taken by many to prove that homosexuality was based in a structural difference in the brain, despite the fact that the INAH neuron set has not been demonstrably connected with any particular behavior. LeVay has been very careful to say that his finding is provocative, but not conclusive. Nevertheless, he was featured on the cover of *Discover* magazine, under the headline "Sex and the Brain: One Brave Scientist Is Exposing the Link."

Something as crucial as sexual preference is a lot to hang on a speck of neurons with no proven function. And since there's good evidence that the hypothalamus, including the INAH group, is identical in newborns, differentiating in boys and girls by about age ten, that leaves a lot of time and room for a host of influences. My purpose, in part, in exploring structural gender differences in the brain is to keep them in perspective: they are few, they are slight; we don't know what causes them, and in many cases we don't know what they do. On the other hand, they are real. Not knowing what they do means that in some ways the brain is as much a mystery to us as it was to Paul Broca.

Given that, there are also a few other neuron variations associated with differences in sexual orientation. Before LeVay's study, Dick Swaab's group in Holland analyzed another neuron set—called the Suprachiasmatic Nucleus—in the front of the hypothalamus. This group is known to help control circadian rhythms (our personal 24-hour time cycle), and may also be cued to an apparent daily cycling of estrogens and androgens, both of which levels appear to rise in the evening. Laura Allen had shown that this nucleus was different in men and women—more bulbous in females, more elongated in males. Swaab's group, again comparing gay men with straight men, found the nucleus to be twice as large in homosexual males as in heterosexual men.

There's one more comparison, this time beyond the hypothalamus. Among the nerve fibers that connect the two hemispheres of the brain is a narrow strip called the anterior commissure, which tends to be slightly larger, around 10 to 12 percent, in both women and homosexual men.

The connection between the hemispheres, as it turns out, is another logical place to look for gender differences. This idea is based on a somewhat controversial theory that, although the brains of men and women look alike, we use them very differently. Specifically, the theory says men are more "lateralized"; they rely on one hemisphere or the other in doing a task. By comparison, in this argument, women use both. This came in part from studies by neuroscientists such as Michael Gazzaniga, who found that male stroke victims suffered much more from damage to one side of the brain, suggesting that female stroke victims could more readily turn to the undamaged hemisphere for help. More recently, researchers at the Institute of Child Development in Minneapolis tested the electrical brain activity of men and women performing a series of tasks. In tests involving word selection, women recorded equal activity in both hemispheres. Most of the men—there are always exceptions—showed increased activity only in the left hemisphere.

Also, studies of women exposed *in utero* to DES (diethylstilbestrol)—a synthetic hormone once thought to prevent miscarriage—produced suggestive results. DES was widely prescribed between the 1940s and 1970s, until studies in rats indicated that it acted as an androgen mimic, masculinizing female rat fetuses. While no such dramatic findings have been produced in humans, a brain-imaging study did indicate that women exposed to DES prenatally might use their brains in a more lateralized manner. In this test, a simple instruction to listen to a series of statements showed that only one hemisphere in the DES-exposed subjects was consistently responsive, whereas in unexposed women, both halves were active.

One of my favorite gender-difference theories came out of this hemisphere-coordination notion. It is a proposed explanation of why women seem to find it easier to talk about emotions: If the left brain handles language and the right brain handles emotions, then women would be better at expressing emotions and pulling everything together, whereas men would be more disconnected. It's such a neat little theory, but it seems to fall into that journalistic category of "a story too good to check": If women really *do* chatter so efficiently back and forth between brain hemispheres, you would also predict a visibly better system of connection in women. (Think of it like a telephone system: there's no need for multiple lines if conversation is rare, but if the various parts of either side are in constant consultation, you'd need more lines.) Those wires haven't been found—at least not enough of them. There is that small difference in the anterior commissure and there are some contradictory studies of the larger web of nerve fibers, known as the corpus callosum, that bridges the two halves. There's also the suggestion—a sort of overall impression—that the corpus callosum is larger in women. But it's not a difference great enough to be noticed at autopsy. The tail end of the corpus callosum generally seems to have a more bulbous shape in women, too. But what that means is anyone's guess. The difference appears in adults, but not in boys and girls. Is it the result of experience? Is it something set by an internal clock, triggered by the release of estrogens in puberty? We don't know that either.

All of this raises a broader question: Can we really expect dead brains, so completely motionless, to tell us about who a man and woman were and how they thought when they were alive? The brain after death is so silent, such a contrast to its lifetime of high-energy communication, that it's not surprising it reveals such unclear results, such wobbly uncertainties. It's still the most precise way to map a structure—counting off cells. But it's deciphering the count that's so difficult.

It's not entirely surprising that so many scientists have turned to studying the living brain. Neither is it surprising that the rare researcher with an assertive theory—like LeVay, linking the INAH-3 cells to a particular, controversial pattern of human behavior—receives such a public embrace. Live brain studies track the flow of blood, the hum of energy within. They've tended, recently, to produce the most dramatic reports of sex differences. But it's important to emphasize that just because we don't understand or can't see the brain's fine structural changes doesn't mean they aren't important. The gap in our understanding of how function flows from form may be an inability to interpret what we see. Yes, we've advanced by breathtaking leaps beyond the scientific horizons of Broca. But our curiosity and impatience still outstrip our knowledge. It's apparent, when you push the point, that a few hundred INAH nerve cells can't be understood by looking at them in isolation.

Estimates of the number of neurons in the human brain run as high as a trillion individual cells. And all the neurons talk to each other, relay information, and network in ways that we are only beginning to understand. Studying brain structure is still like peering down at Manhattan through an airplane window at night, seeing that fantastical rise and fall of light, and trying to know—from that distance, in the dark—who lives there, who's talking to whom, and what they might be saying. We expect so much of neuroscientists because they've shown us so much, but any one of them will tell you that the brain does not yield its secrets easily. "We used to think we could put it all together in easy pieces," recalls neuroanatomist David Amaral. "That might have been the conceit twenty years ago. But now we're more respectful."

And, he adds, "you can't judge the importance of neurons by the number." As an example, he cites the jittery noradrenergic neurons, which send out alarm signals. They are found mostly in two very tiny clusters, numbering less than 100,000 neurons. Yet each one of those neurons, Amaral points out, can contact between 100,000 and 200,000 additional neurons, so that their influence —across billions of neurons—is enormous. That's one reason why our response to danger is so quick and so powerful: even after a threat vanishes, our lungs still gasp and our hearts still pound.

Marc Breedlove at the University of California, Berkeley, puts his argument this way: "I believe that all differences in behavior are based in differences in structure. Otherwise, you have to believe in ghosts and spirits in the brain. There's no doubt, when you look closely, that there are structural differences between the brains of men and women. The whole question is, how does the brain respond to those differences?"

We've all heard that girls mature faster than boys, that they talk and move gracefully at an earlier age. It turns out this is not merely parental chitchat; there may be a neurobiological reason. The brains of female and male appear to develop along different tracks, although no one is sure why. Harvard psychologist Jerome Kagan calls it one of the two real mysteries of human evolution. (The other, he says, is why women outlive men.) In other words, sex differences are not constant. The gaps may widen and close according to our age. A young girl may use words more easily than a young boy, but in later years he's likely to catch up.

Some of the best evidence that difference in development is not simply a matter of culture, a result of the way we educate children, comes from looking at similar behaviors in other animals. William Overman, of the University of North Carolina, at Wilmington, directly compared very young humans, aged 15 to 29 months, and infant monkeys on a skill test. The test involved a set of eight objects (such as colored blocks) arranged in pairs. One pair, for example, red star-shapes, always included a food treat. Overman wanted to know how quickly the youngsters would make that association. Consistently, girls learned first. The same was true for monkeys—small females went directly to the food much faster than males. However, if Overman then switched the treat from set to set, the males, both human and monkey, were much quicker to expand their search. The females found it harder to abandon the known answer.

Wellesley College neuroscientist Joanne Berger-Sweeney and colleagues also compared sex abilities, this time in very young rats. Berger-Sweeney is interested in nervous system communication using the neurotransmitter acetylcholine, which carries impulses from one nerve cell to another. Acetylcholine is a derivative of choline, part of the vitamin B complex. It is active in the nervous system throughout the body and, she suspects, helps build the young brain, particularly in regions linked to spatial reasoning.

Working with newborn rats, she decided to find out what happened if the animals were denied the neurotransmitter. Berger-Sweeney and her colleagues deliberately damaged the pathway for acetylcholine. When the rats were a few weeks old, she tested them on a maze, probing for effect on spatial navigation. Unexpectedly, she found that the males were fine, but almost a third of the females wandered in meaningless circles. Her theory was that, at birth, females were making a lot of acetylcholine, industriously building up needed connections. The lesion was therefore extremely destructive. In males, there was little effect; they hadn't booted up that system yet.

She decided to test it another way. This time, she blocked development of the cholinergic system in rats, between one day and three weeks old. Several weeks later, she tested that group. This time, the females—who had their basic system in place—were fine. But the males were baffled.

"Female and male brains do appear to develop at different rates," Berger-Sweeney says. "I don't think you can say anything generic. It's not that female brains mature, overall, faster than male brains; it's that different parts of the brain grow at different times, at different rates, depending on whether you are male or female. And why is that? It's something we're still trying to understand." One of the questions that Overman raises is whether male and female brains are ever on exactly the same development curve: "Maybe everybody's brains equalize as we grow up or maybe they don't."

Raquel and Ruben Gur, married to each other and partners in neurobiology at the University of Pennsylvania, have found a similar pattern in aging human brains. They recently reported that although men have bigger brains on average, they also lose brain tissue at almost three times the rate of women as they age from their late teens until their mid-40s. And this appears to be caused by the relative efficiency, or lack of it, with which the respective sexes use the brain cells they have left.

"The picture you get is that women seem to be able to reduce the rate of neuronal activity in proportion to the [brain] tissue that they lose, whereas men continue to overdrive their neurons," says Ruben Gur. "Even though women, too, lose tissue as they age, they seem to be riding herd, keeping the system more level." It's a paradox of sorts, because you might think that high energy use of neurons would be more productive. But it turns out that as the brain ages, it needs to conserve energy more. Failure to do so may have an unfortunate result; overdriving can cause cell death. When cells are working too hard, they generate wastes that can actually injure them and other nearby cells.

The differential loss of brain tissue with age between men and women is most visible in the frontal lobe, Gur says, where the brain handles planning and inhibition. "Men lose tissue in the frontal lobe at such a rate that by the time they reach middle age, even though they start with larger brains, their frontal lobe is the same size as the frontal lobe in women."

When I began to question the importance of tiny structural variations in gender differences, the Gurs were among the first people I called. Since the early 1980s, they've produced a remarkable set of studies that analyzes both brain structures and the activities that those structures support. More recently, their work has focused on the live brain, using imaging technologies to study differences in the way men and women respond to the same situation. I wondered if they had grown less enchanted with structural differences —if they'd decided that we could only interpret male and female differences through live brain work. Is neuron counting a futile exercise, or does it ultimately mean something about who we are? After consulting with his partner, Ruben Gur put it this way: "It is impossible to understand brain-behavior relationships by looking at either structure or function alone. And this is probably true for all of biology. The main point is that structure and function are intertwined, and we need to study both. And so, no, we don't think that we (the research community) put too much emphasis on neuron counting."

It's fair to say that the function studies, at least recently, have been by far the most compelling. The Gurs use two modern technologies in studying the brain—Positron Emission Technology, best known as PET scanning, and Magnetic Resonance Imaging, or MRI. The classic PET study involves injecting a sugar solution—technically, glucose—into a volunteer. The glucose carries a radioactive atom, which releases a signal that can be measured by the scanner. When cells are working very hard, burning energy, they suck up more glucose to sustain them. So if you want to know what part of the brain responds to a particular challenge, you can use the PET technology to see which regions light up, indicating a heightened metabolism. MRI, by contrast, uses a magnetic field to induce hydrogen atoms to spin within living cells. As the atoms slow down again, they release a charge specific to the individual cell—nerve, muscle, blood—which can provide a very precise picture. The Gurs used that technique to calculate the volume of neurons in the frontal lobes of the brain.

The Gurs have used PET to study how the brain divides its resources, where it uses energy in a certain task, where it relaxes. In the past decade, the Gurs have shown that women's brains seem to run a little hotter, circulating more blood, and maintaining a little more overall charge. You could argue—although they certainly do not—that here too might lie a possible explanation for the overall brain size variation: that perhaps women use their brains more efficiently; they don't need a bigger brain.

A number of recent live brain studies, mostly using PET scanning techniques, have shown that men and women seem to activate different parts of their brains while doing word tasks or math/spatial tests. There's an ongoing debate over these results: Are they created by what we are taught, or are they real gender differences based in an inborn biology? Are women naturally better at verbal tasks? Are men born talented at map reading and geometric calculation?

Remember, these questions refer to average behaviors. They are not individual predictions. Many of us know math-savvy girls and boys who can't make it through calculus. But if these overall differences are real, have they been handed down, generation by generation, through genes, or culture, or some combination of both? If girls tend to do less well at math, is that because it naturally comes harder for them, or because teachers expect them to have little talent for it, thus lowering the girls' expectations of themselves?

There is, actually, a quite solid evolutionary argument for a spatial reasoning difference between the sexes. It's supported by some good research work using both animals and humans in mazes, but not by any clearly visible difference in human brains (sigh). The evolutionary argument is based—aren't they all?—in pure sex. That is, in polygamous species where the males wander about and the females stay near the nest, it is the males who most need to keep maps in their heads. The theory would predict that in such species, the hippocampus—the structure in the brain that is thought to organize spatial analysis—should be bigger in males in response to all those expeditions and resulting mental maps, and that those males should be better than females at negotiating one of those tortuous laboratory mazes that scientists love to dream up.

Meadow voles, those fluffy and promiscuous little rodents, fit the theory perfectly. The male has a larger hippocampus. And he is a whiz at mazes, especially compared to the female. Now compare these to another species, the monogamous prairie vole, whose males and females both stay home. Among prairie voles you find no significant gender difference in either hippocampus or ability to navigate a maze.

But what if the need to wander was reversed? What if it was the female who needed to be out there traversing the unknown? You can pursue the answer to that among cowbirds. Brownheaded cowbirds are nest parasites. They don't rear their own young; the females sneak their eggs into the nests of others and let the other species' moms do all the work. To do that, the cowbird female has to make a map in her brain as she prowls in search of the best nest. If you compare the hippocampus by gender in cowbirds, you find that the females consistently develop the larger structure.

My favorite part of this is the implication that the wonders of human math/spatial skills are based in sexual promiscuity. From the perspective of American culture, which tends too often to regard math as one of two extremes—a terrifying impenetrable jungle or a cure for insomnia—this at least offers a new way to view the subject. Still, if you accept that in polygamous species males develop stronger spatial skills, we should also acknowledge that most anthropologists believe our species has at least a mildly polygamous history. Based on that, the prediction would be for human males to possess slightly stronger spatial skills. Men do tend, overall, to outperform women on math/spatial tasks. But no one has found anything like the distinct sex differences in the hippocampus that are seen in meadow voles and brownheaded cowbirds.

The only study so far that has effectively compared the hippocampus in men and women looked at another aspect. Research presented at the 1995 National Society for Neuroscience meeting showed that the human male hippocampus apparently ages faster than the female's. Men lose cells more rapidly—in a pattern much like that seen by the Gurs in the frontal lobes. There is also a correlated effect on spatial ability: young men do better on such calculations than older men.

The speculation in that study was that the aging difference might be directed by androgens. But steroid hormones aren't the only possible agents. As Berger-Sweeney and others have shown, there are many systems affecting brain development. Still, the rodent studies are remarkably crisp in showing that androgens, notably testosterone, are linked to spatial reasoning.

Most male rats—also an aggressively polygamous group—do better than most females in maze tests. But you can change that easily. Among rats, sex-typical differences in maze performance can be completely reversed by early administration of the right hormone. If a female gets a solid dose of testosterone at a very young age, she solves mazes like a pro for the rest of her life. In humans, of course, you can't do that kind of test. But as our adventures in sexually ambiguous bodies showed in the last chapter, natural hormone disruptions do occur. And they provide circumstantial evidence that, yes, testosterone makes a difference. For example, human males who are testosterone-blocked—those who lack the important androgen receptors—tend to have much lower math/spatial scores. By contrast, females who have unusually high androgen levels—such as those with Congenital Adrenal Hypoplasia—tend to score higher in math tests than females with usual androgen levels, although these results are somewhat inconsistent.

But beyond that, many, many studies show that men are more comfortable with maps; they do better at spatial rotations in their heads and they consistently produce the highest of high-end math scores. Women show up at the high end, too—just not as often at that ultimate peak.

A very neat set of studies, pulling together animal and human observations, was conducted by Thomas Bever at the University of Rochester, New York. In a rat maze which included landmarks, he found that female rats used the landmarks to negotiate. If he removed the landmarks, such as bright-colored objects, leaving smooth, blank maze walls, the female rats were lost. Removing the landmarks made no difference to most of the male rats. They seemed to navigate by calculating distances traveled. If Bever changed the length of a maze arm but left the bright landmarks, the females, temporarily, could outperform the males.

He found a remarkably similar pattern in humans. Bever set college students the task of navigating a computer maze, nicely cued with landmarks. Once again, he found that changing the geometry threw the men; altering or removing landmarks threw the women. Everyday observation seems to support this. Think of the way women give directions, compared to the way men tend to do it. Washington University researcher Steven Petersen put it in personal terms: "My wife and I are completely different on this one," he said. "If we have to tell someone how to get to, say, the library, she does landmarks—the furniture store, the gas station—whereas I say, go a mile or so on this street, turn west, like that."

It's clearly worth raising the question of whether this behavior is rooted in our biology. Does testosterone lurk behind these kinds of differences, nourishing whatever circuitry likes mathematical games and mental map-making? Is it indeed a predisposition that comes with early "masculinization" of the brain? There's evidence to suggest so. Our early history was polygamous. Today we're a far from perfectly monogamous species. Let's concede that testosterone may, in some poorly understood fashion, polish map-making for males. But let's also concede that culture may have helped maintain and reinforce that, with its legacy of assuming male superiority.

For instance, Jacquelynne Eccles of the University of Michigan has shown that parents tend to reinforce such gender stereotypes. Interviews with mothers and fathers of American boys and girls, especially those doing well in math class, found that the parents attributed the boys' success to natural talent and the girls' success to hard work. (The same attitude split, incidentally, occurs in the way parents assess their children's athletic achievements.) Also, Eccles says, if you ask parents which skills are important for their children to learn, a majority will say that boys need math, computer science, or mechanical skills. They think girls should focus on English and biology.

Culture adds complications. In other species, it's much simpler to link an ability to testosterone. That's not saying hormones don't influence us as well; human behavior is certainly biology-based at some level. But if a girl is born very talented in math/spatial skills and yet is actively discouraged from exercising that talent at every turn by the people most responsible for nurturing her, how likely is she to develop her ability? The answer to that might involve other personality traits—competitiveness, stubbornness, mental stamina—also perhaps based in biology, also certainly influenced by culture.

How do we sort it all out? The danger of leaning too hard on biological explanations is that they turn into excuses. They can be used to argue against equal opportunity. You can imagine the argument: Why waste precious time and educational resources teaching girls math or making engineering careers available to women if they'll never be as good in those areas as the boys and men? It's too easy to fall into prejudicial thinking.

Richard Lynn, of Belfast, reminded me of this in a transatlantic telephone conversation as he argued the spatial superiority of men. He described playing chess with his daughter and her inevitable series of losses. Lynn explained that he never expected her to win —the math-spatial handicap, you know. On the other hand—and by this time in the conversation, I had broken a pencil—she speaks French well; the female verbal ability, you know.

In fact, there's another complicating issue here and that is the matter of perception. It may not be that men are necessarily better at math-spatial tasks than women; it may be that they simply handle those requirements differently, as illustrated by Bever's rat and student experiments. Since we live in a culture that is still male-dominated, female performance is held to a male standard. If you accept that women "do it" differently—say, landmarks over geometry—it's not necessarily a question of better or worse. Perhaps, some scientists suggest, the sexes mentally organize the world differently. Men automatically map it out, women make mental maps only as needed. That would suggest less a difference in the ability to use maps than in the propensity to create them. Think of it like this: I drive my husband to a new doctor's office, and, relaxing in the passenger seat, he almost absently maps out the route so that, with very little help, he can navigate there again. He drives me, and I admire gardens, critique the paint colors on houses, watch the faces of people at bus stops. Only if I know that I might have to drive back there myself do I have a vague sense that we went, sort of, that way. Once I drive myself, I do carefully inventory turns and twists and I can usually find the office again without too much trouble.

There is, though, some consistency to studies finding that boys do better in math-based schoolwork and girls possess a verbal edge. The most comprehensive study of test differences, analyzing 32 years of scores, was published in *Science* in 1995. University of Chicago researchers found these results: In math and science, boys in the top ten percent outnumbered girls three to one. In the top one percent, there were seven boys for every one girl. In some mechanical-vocational tests, such as electronics and auto repair, there were no girls in the top three percent.

If, by contrast, word skills were tested, girls looked startlingly better. There were twice as many boys as girls at the very bottom of the test scores, and twice as many girls at the top. Reading comprehension tests were dominated by girls; in writing skills, girls were so much better that the researchers warned that "the data imply that males are, on average, at a rather profound disadvantage in the performance of this basic skill." The study pointed out that while some differences could be explained by culturally influenced curriculum choices—boys tend to take electronics, girls don't—that can't be true with writing, which is required of all. We could, in fact, dismiss Richard Lynn's theory—that men's brains are larger to accommodate superior spatial abilities—by making the equally improbable suggestion that men communicate less well because the math-spatial chunk has taken up too much room in men's brains, crowding out verbal processing. But there's a more serious issue here. Much has been made of the way schools fail girls, but, clearly, after surveying these test results, we must consider that they are failing boys as well. If, indeed, we accept some biological gender differences in math and word skills, perhaps our aim should be a curriculum that recognizes the strengths and weaknesses of both sexes.

As with spatial ability, there's an evolutionary argument for females as stronger communicators. We can also ground this in ancient polygamous society: While the males were out wandering, searching out mates and running each other off, the females were close to home, raising a family. They were reassuring the young, teaching them how to survive—a job requiring strong communication skills. But women also had to travel around gathering food, which would require some map-making skills. And I find it hard to believe that language development was so separate in men and women that the ancestral females were spinning complex conversations while the ancestral males were still out there shouting "Ugh." After all, half of those youngsters the mothers were talking to, and teaching to talk, were boys.

However, there are some interesting structural differences in male and female brains that relate to this topic. Most of us pick words out of noise better with the right ear (and thus the left brain). But, in general, women show less discrepancy between the ears; they are more tuned to listening with both. In the 1970s, researchers reported that a part of the temporal lobe, the planum temporale—thought to be important in speech processing—was more equal from left to right in women than in men. In men, it tended to be asymmetrical, suggesting heavy use by the left brain. This suggests that the two sexes are processing language slightly differently, in both the way they listen and they way they talk.

There's also some support from one of those nitpicky cell-counting studies. At the 1995 meeting of the Society for Neuroscience, Sandra Witelson, of McMaster University in Toronto, reported that women have a greater density of neurons in the temporal cortex, part of the cerebral cortex. By comparing dead brains, Witelson and her colleagues were able to probe systematically through six layers of neurons in the cerebral cortex. Toward the bottom, in layers four and five, they found that women had about 15 percent more neurons, packed more tightly into the same space. Take note: They didn't count up the whole cortex. They took pinhead-size samples from each layer, considering them representative. The samples for layers four and five contained about 35,000 cells in women, about 5,000 less in men. These layers are linked to language—particularly in the left hemisphere—and recognition of melody and tonal quality of speech in the right. Witelson suggested this density difference may explain test results, especially why girls learn to read faster and why women outperform men, overall, on simple alphabet tests.

The Gurs' live-brain work also hints at an affinity women have for words. In early 1995, they did an unusual study in which they focused not on the brain at work, but at rest. They asked a group of men and women to simply relax, let their thoughts wander anywhere (except into that relaxed state known as sleep). They then monitored basic brain metabolism, injecting glucose and using the PET scanner to see where in the brain it was being absorbed most quickly. I tend to wonder, a little, where these thoughts traveled. Do all of us relax equally? In my one yoga class, when we were told to relax by envisioning a lovely pond surrounded by trees, I kept thinking about mosquitoes and water snakes (I grew up in Louisiana). It's certainly possible that today's culture might separate even minds-at-rest into some stereotypical split, wherein she's thinking about what to fix the kids for dinner and he's wondering why his boss has to be such a jerk. Nevertheless, the Gurs produced brain images that were remarkably alike, except in one small and curious place—in the limbic system. The word "system" refers to a complex structure, down in the back of the brain, which controls many of our split-second reactions and is also thought to provide the neural foundation for emotional processing.

Within this system, men's brains glowed most brightly in a region linked to a quick physical response. Women's didn't; their limbic system was more active in another region, linked to a quick verbal response. Equally interesting, the strong male response sprang from a part of the system considered evolutionarily very old; it also exists in reptiles. But the female response derived from a region called the cingulate gyrus, which appears more recently in evolutionary development and is unusually large in humans. Ruben Gur speculated that this suggested a fundamental physiological difference that would lead women to respond first with words, men with fists.

About the same time, another married neuroscience team, Sally and Bennett Shaywitz, of Yale University, also published a report on gender differences in the brain, this one focused specifically on language processing. The Shaywitzes asked their test subjects to do a rhyming task while the researchers made detailed MRI images of their brains. Their work gave additional credibility to the theory that women draw more readily on both sides of the brain. In most of the men, as they matched words with appropriate rhymes (cake, bake), a small center called the inferior frontal gyrus (behind the eyebrow) lit up only on the left side of the brain. In women overall, the region lit in both left and right hemispheres.

On the surface, these studies of function seem much more impressive than the structure work. Instead of trying to sort out a shape difference in a piece of the brain about the size of a sugar grain, we see men and women performing differently. The contrasting images produced by the Shaywitz laboratory were truly startling and, not surprisingly, the prestigious British research journal *Nature* printed them on its cover. Yet, in their own way, these results are also difficult to see in perspective. Maybe men and women are taught to process language in different ways, just as there could be culture-influenced differences in the way we let our minds go "blank."

Perhaps girls are encouraged to use words more and taught to communicate differently, so that a woman's brain would approach a task in a different manner than a man's would. Are the Shaywitzes' results showing a difference in inherent biology, or a cultural difference, or the interplay between them? When I asked the Gurs about this, Ruben (again consulting with Raquel) answered: "Here, certainly, we should refrain from pretending that we have an answer. But we do think it is safe to stay away from either extreme position [all biology or all culture].

"Even as we are struck by the similarities in the behavior of identical twins," he continued, "there are equally striking differences. The same biological constitution may lead an individual in one culture to become a crime boss, in another, a college dean. On the other hand, it doesn't make sense to us that at some unknown point in time, all cultures on this planet decided that one set of behaviors is appropriate for males and another for females. And that since then we have all followed these classifications blindly—even if they go against our nature."

The neuroimaging studies, the Gurs add, tend to show that a biological difference will predict a behavior. They see that in the limbic system comparison. It lets them predict, Gur says, that men are more likely to strike out physically and women to strike out verbally. And that is pretty much the way people tend to behave. "In these cases, it's tempting to conclude that it's biology, not culture, that produces the behavior," he adds.

But let's consider another neuroimaging study, very different in its findings: Also in early 1995, a team from Washington University in St. Louis published a comparison of the way men and women handle word tasks. Theirs was a word association game: "Cake goes with . . . ?" There was no wrong answer; it could be "birthday," "fat," "frosting," or "chocolate"; it just had to associate with the first word. Unlike the Shaywitzes' study, the St. Louis group, led by psychologist Steven Petersen, found no significant differences between the way men and women processed language. Although men's brains seemed to be working a little harder (blood flow was a little faster, response a split second slower), both sexes activated the same part of the prefrontal cortex.

Petersen points out wryly that his study received little public attention, unlike the Shaywitz and Gur studies, which made the cover of national news magazines. He worries that our fascination with gender differences misses the main point, that we are mostly alike. And, he argues, finding a functional difference tells us as much about culture as it does about biology. Remember his earlier example of the way he and his wife give directions to the library? In that case, Petersen says, "It wouldn't be surprising to find that different parts of the brain engaged for men and women. And it might only mean that, in our culture, we'd been taught different ways of giving directions."

Now, bear with me as I find my way back (with no landmarks) to the autopsy room at the beginning of this chapter. Petersen's point is that our brains are basically alike, even if we use them differently. If we set the two brains side by side again, they look the same. We have to work hard to find what's different. The contrasts are too tiny, and still far too mysterious, to suggest that these are profoundly different organs. They may do some things in different ways, but the basic repertoire is the same: I can tell you how to get to the library as well as the next guy.

Chapter Three

HEART TO HEART

Sex Differences in Emotion

The heart is mostly muscle. Its steady thump—for all that we invest it with the rhythm of romance ("It beats for you, darling")—is merely the sound of machinery at work, pushing blood where it needs to go. Only in a too narrow window of childhood do we believe in the heart as a literal container, a dispenser of love. When my son Marcus was four, he asked one morning before our daily separation—I to the office, he to preschool—if I remembered him when he was away. "You're always in my heart," I replied, smiling. He looked back at me, very sober. "Will you keep me in the point?" he asked. "I want to be there." And he wanted his daddy, he said, in the upper curve. For him, the heart, my heart, was a valentine-shaped container to hold him, his father, his beloved cat, and a world of security. He's left that behind now, as any properly cynical seven-year-old would. And me? Not a chance. I still hold him in the point of my heart.

While it's not rational, it's unexpectedly hard to demystify the heart, to let it be what it is: atria, ventricles, and valves packed in cardiac muscle, pumping some 9,000 liters of blood a day, in and out, thump and thump, pacing the biological rhythms of our days. Mine beats—in all technical accuracy—for me alone. Yours, for you. We all know it; it's the ABC stuff of high school biology. And yet we still have heartthrobs, hearts and flowers, heartbreaking songs, heartfelt, follow your heart—as if the brain is all cold logic and the heart all warm feelings.

Maybe there is actually a scientific reason for this. One theory, at least, links the heart and emotion in a most logical way. It goes back to the developing fetus—the flickering start of a human being—floating in the amniotic sac, where the loudest, steadiest, most reliable sound is the beat of the mother's heart. We know that somewhere between five and six months' gestation, the fetus begins to hear. That world of darkness and warmth is not silent; somewhere between 50 and 160 times a minute, the drumbeat of the heart repeats, as constant as time. And after birth—into a world shockingly full of noise and light and dry, rasping air—the baby can still hear that reassuring rhythm if the mother holds her (or him) just so. The sound that so recently defined safety is still there, still close.

It's a fact that mothers preferentially carry infants on the left hip. It's true for right-handed mothers, who rationally explain that it makes sense to free their more useful hand for other chores. But it's also true for left-handed mothers, who rationally explain that it makes sense to use their stronger, dominant arm in holding the child safely. For women, it has been true across culture and history. It likely grows out of our evolution: Consider that ape females, too, prefer to cuddle their babies to the left. Perhaps in some distant past mothers learned that their infants settled and relaxed if they were held close to the heart.

Could each new mother—ape and human—learn this lesson anew? Not likely. If so, why wouldn't the left-carrying bias be true for fathers? It's not nearly as strong, not even in fathers who spend a lot of time caring for their small children. Mostly right-handed, as is the human species at large, they're more likely to hold the child in the stronger arm, across the body from the heart. It makes rational sense, but it apparently runs counter to an emotional-biological sense built into the mother-child bond. In that crucial, virtually universal relationship, the heart really is more than a pump. It is a signal. It binds us, giving the small child his first sense of security in the world outside the womb.

Frans de Waal, a primatologist and professor of psychology at Emory University in Atlanta, comments that people sometimes try too hard to prove that they are making an intelligent choice. Is every action of parenting based on intellect rather than (if you will) the heart? Unlikely, says de Waal, who proposes an open-minded attitude toward the influence of biology. We come from a species—a series of species, really—with a long history of females as mothers, nurturers, and supporters. There's a graceful, evolutionary kind of logic to the way a mother cradles her child, and, maybe, an equal logic to the distance a father leaves between baby and heart. That's not ever to deny love between father and child. In either case, we're talking about a parent carrying a child, holding a baby close, and providing security and shelter. The width of a human body is not large, but the space is still curiously suggestive.

Of all the differences between men and women, the one that seems currently to grab the most attention is that of emotional connection. It has become the stuff of stereotypes and greeting cards: women seek commitment and men flee; women talk about their feelings, men change the subject to basketball scores; women share their emotions with friends, men regard that as an act of indecent exposure. Is it merely perception? And even if it is, where does that perception come from? How soon does it separate the boys from the girls?

De Waal pays tribute to the work of New York University psychologist Martin Hoffman, who has studied the emotional responses of day-old infants too young for anyone to "train"—before any suspicion of socialization. Hoffman simply let the babies listen to sounds—of other babies crying, of animal calls, of the weird droning voice of computer-generated language. The babies responded most strongly to the sound of another human in distress. But it was the tiny females who reacted most intensely to the sound of another's trouble—a reaction that Hoffman suggested would run like an underground stream through their entire lives. His inference was that, while both sexes respond to another's distress, even on the first day girls are more tuned to an empathetic response.

Hoffman's carefully technical analysis puts it like this: "Females may be more apt to imagine how it would feel if the stimuli impinging on the other were impinging on the self."

Females, for so long the first line of caregiving—the first defense, really, that an infant has against the world—*have* to be oriented to the needs of others. No wonder, then, that females possess such exquisitely tuned senses: The female sense of smell is more acute than the male's (especially during ovulation), and women are more sensitive to touch than men. As it turns out, ability to communicate with touch is critical for the healthy survival of a child. Studies with premature infants have found that if they are held, even just gently stroked, they grow and mature faster. They gain weight 47 percent faster than those left alone, even if both receive the same amount of food.

Touch now appears to be one of those factors that the genes rely on to guide proper development. In some animals—rats, for instance—if the mother isn't there to cuddle and lick, the babies simply don't make the hormonal chemistry necessary to grow. Saul Schanberg, at Duke University Medical Center, suggests that because mammals are so dependent on a mother's care, her touch sends a signal of reassurance that the world is all right. Denied it, even for 45 minutes, a newborn rat's metabolism slows in a seemingly self-protective measure, shutting down requirements for nourishment.

There's another, almost charming, study of rhesus macaques—an agile species of Asian monkey—in which University of Wisconsin scientists found that the mother's holding an infant near the chest directly influenced what we sometimes think of as the chemistry of happiness, raising the natural opiates of the bloodstream to the point that the scientists compared the pleasure of touch to a drug-induced high.

And then there's hearing. Females hear high-pitched sounds better than males. That difference begins in childhood and actually becomes more acute. In one study of simple noise, a researcher asked 24 men and 24 women to turn up a speaker until the wail of sound just passed the comfort zone. The men averaged eight decibels louder than the women. Men are comfortable with sounds twice as loud as women are. This is equally true for boys and girls even as young as age five. Examining the tiny hair cells in the inner ear that vibrate to transmit sound, researchers have found women's ears to be more sensitive, men's less. The number of hairs are about the same, but women's vibrate more intensely—and the richer and more complex the vibrations, the better the hearing. The exception is women who have (or had) a twin brother: Their ears are also less sharply tuned. Scientists suspect some masculinizing influence of early development side by side with a male—perhaps an extra wash of testosterone in the amniotic fluid.

Mothers have to communicate with children—not just comfort them, but warn them of danger. This protects the individual child and may, in an expanded picture, be essential to the overall survival of the species. Researchers regard such communication—nurturing, protective—as so necessary that they suggest it created women's verbal advantage over men.

Evolutionary psychologists such as Anne Fernald at Stanford University have found that mothers talk to their infants in a particularly high, crooning voice. A mother's voice goes up by as much as two octaves, the equivalent of 16 ivory-colored piano keys, when addressing a baby. Fathers' voices rise also, but less dramatically—perhaps half an octave, or four piano keys. Fernald finds the same maternal pattern across cultures. American, French, German, Italian, Chinese, African, Japanese—all mothers scale up for infants, and not just for their own. Babies respond to the tone more than the words. An American infant smiles as readily at the music of a Japanese voice as at that of an English one. It's the "song" that counts. (Recently, British scientists suggested that this maternal music fits beautifully with the idea that carrying babies on the left helps nurture them emotionally. The position means that a mother would be crooning into the infant's left ear. Thus the sounds are processed by the right hemisphere of the brain, where neuroscientists think emotions are mainly processed.)

"I'm arguing that the vocalizations of mothers are well matched to the sensitivities and needs of the infants," Fernald says. At the same time, she points out, this style of communication would be distinctly out of place elsewhere. Imagine turning to a coworker, or a fellow subway passenger, and talking as we do to a baby: up close, in the face, suddenly high in the fluting notes of the treble range. "They'd think you'd gone crazy."

Fernald has found, by playing tape-recorded voices to them, that infants turn more readily toward a woman's high-note sounds than toward the tones the same woman might use to an adult. And the heart of the infant slows, calms, and steadies, beating more gently as it hears that particular music of a mother's voice. Comfort, in this sense, is anything but a luxury; the comfort that mother gives child appears basic, biological, and continuous with (perhaps part of) the same developmental process that goes on in the womb. It is in this intertwining of mother and child—a tale of two hearts—that the evolutionary argument for emotional differences between men and women begins.

We don't, of course, possess a time machine—we can't observe ancient behaviors, "prove" a female history of emotional connection. Evidence of past behavior is circumstantial if it's there at all. Scientists use fossilized bones, primitive tools, and long-abandoned campsites to create a picture of ancient societies. They analyze our current biology and make highly educated guesses about what shaped it. Take, for example, the theory of Margie Profet, an evolutionary psychologist at the University of Washington-Seattle, that morning sickness—that hateful and apparently useless aspect of pregnancy—derives from a past in which our ancestors had not learned everything about what was poisonous and what was not. Thus a built-in nausea and a sensitivity to strong-flavored foods, especially in the first trimester, might well be very protective of a human embryo. We inherit the behavior, essentially, out of our past.

To avoid cultural biases, scientists also study sex differences in other animals. They explore the biology of other families and bonds and coalitions. The field of comparative biology is a fascinating one, but needs to be approached with caution. No animal makes a perfect match for human society, just as our behavior could never fully explain the social hierarchy of wolves or the lifelong partnerships of Canada geese. But as fellow species on the home planet, we also share certain characteristics. Like wolves, we have a social hierarchy; like geese, we build partnerships. So scientists look where they can for comparisons and study a range of species to try to create a mosaiclike impression of behavior. The animal considered, usually, to provide the best comparison to human behaviors is a fellow primate—the chimpanzee. This great ape is highly intelligent, highly social, and a bare sliver of DNA away from us. Geneticists estimate that we share some 98 percent of our genetic material with the common chimpanzee and perhaps one-half percent more with the pygmy chimpanzee, also called a bonobo.

Looking at the common chimp for now, let's note first that its relationships—especially concerning mating and parenting—are not like ours. Chimpanzees are an enthusiastically polygamous species. The males do not contribute, directly, to raising the young. They're friendly to the youngsters; they may occasionally tease and play with them. They also give females food, sometimes in exchange for sex, which the females then share with the young. The males' size and strength may help protect the group from predators. But that's about as close as chimpanzee males get to involved fatherhood.

And there's no observational evidence of a long-lasting emotional bond between adult males and adult females. Despite these differences from our society, anthropologists suggest that the chimpanzee system may provide clues to our past, to the behavior of our ancestors. The hominid line—our ancestors—split from the chimpanzee evolutionary line only about five million years ago, so one might predict that early human ancestors behaved in ways similar to early chimpanzees. Or even that earlier, primitive hominids behaved something like today's chimpanzees. Many evolutionary scientists believe we derive from a polygamous, chimpanzee-like species. And there are some aspects of chimpanzee behavior that seem startlingly comparable to the way we act today.

As Frans de Waal points out, male and female chimpanzees both form coalitions. The male network, though, is markedly different from the one built by females. The male coalition tends to revolve around hierarchy and dominance. Among male chimpanzees, there's constant negotiation over who gets the power, who becomes alpha male (the chimpanzee equivalent of king), and how long he stays there. Males are constantly plotting coups and negotiating for better allies. In a search for dominance, male A may buddy up with male B one year, then drop him for male C the next, if both agree to make a power play. That doesn't stop them from enjoying each other's company. They play together, tease each other, even (to stretch the analogy to humans) act like friends. They hug each other, carefully groom each other. (This is unlike baboon males, who also form power coalitions but are far too suspicious of each other to touch in "friendship.") Nevertheless, a male chimp will drop a "friend" for pragmatic reasons—competitive success being a major one. Older chimps plot power strategy, baiting ambitious youngsters into competing with each other. Does this sound familiar?

Of course, there's a payoff to all of this, beyond being the chimp who gets the best food and the most comfortable branch of the tree. The top chimpanzee, Mr. Alpha himself, also has more sexual access to females (emphasis on the plural) and is, therefore, more likely than his lowly companions to see his genes continue into the next generation.

As mentioned earlier, there's no comparable reproductive payoff for females. "What's really most important is that males can increase their reproductive output by mating with multiple females, whereas for a female, it does not matter how many males she mates. She can only get one baby at a time. That's why males try to mate with every female in sight, and females are very picky about with whom they mate," de Waal says.

A female chimpanzee's life becomes focused on friends and family, on caring for her young. Chimp mothers nurse for four years and often carry their young for longer, up to six years. It's intensive work and to do it females don't need battle strategies. They need a support system. Their coalitions are made of a close circle of female friends and daughters and cousins and the infants of all.

Within that circle, older females take care of each other and their offspring. These relationships are extraordinarily stable; de Waal notes that some, observed at Holland's Arnhem Zoo, have endured for more than two decades. The females hold the baby chimps, groom each other, sit together, share food. It's almost a perfectly charming picture but, as de Waal has also discovered, even among chimpanzees such codependency has a downside.

Chimpanzee males are comfortable with changing relationships; they don't count on stability in a system that de Waal describes as based on Machiavellian strategies. They always recognize the potential danger in each other. De Waal himself observed two ambitious young males ambush and injure the dominant chimpanzee so severely that he died. That was clearly a coup. After most fights, the males are quick to make peace. They begin to groom and relax with each other.

Chimpanzee females don't kill each other or even hurt each other very often; the aggression rate among females is about one-twentieth of that seen in males. (During territory disputes between two chimpanzee troops, males have been known to slaughter, without provocation, females from the opposing camp. Jane Goodall has speculated this may be a kind of psychological warfare, killing to terrorize. No scientist has observed female chimpanzees ganging up to kill a male.) But that doesn't mean they are perfect models of kindness.

Females don't really fight over access to power, at least not in the sense of seeking pure dominance. Their quarrels usually have to do with access to food or the protection of the young against others. Sometimes there's dislike or even intense rivalry over a particular relationship. And if another female doesn't come through —eats more than her share of the food, or even inadvertently endangers a youngster—de Waal describes the females as "vindictive and irreconcilable." They're angry for days, or even longer; they slap and push and won't come to an ex-friend's aid. And, while it's uncommon, females do appear sometimes to completely lose their tempers, fighting to crippling injury or worse. A trespass, a failure of a trusted friendship, is apparently much harder for females to forgive.

The same appears true in other primates. In one deliberate stretching of tensions in rhesus macaques, de Waal began by playing to the monkeys' love of apples. (Rhesus will stuff apple slices in their mouths until their cheeks puff.) He tossed a single slice at a time into the middle of a hungry group, watching to see who grabbed it first and how the winner was treated.

Rhesus monkeys live in a strong female-dominated society. Like chimpanzees, the females form a close-knit social network dedicated to child raising and protective alliances. But unlike chimpanzees, female rhesus have effectively used those alliances to become the more powerful sex. So, in the rush for apples, the females were the grabbiest. They were also the worst losers. After the scramble for the treat, males were more prone to accept that, well, someone has to lose. In almost half the fights between males, friendly contact was reestablished within half an hour. With females, this happened only 18 percent of the time. It may be, de Waal notes, that females simply reconcile in less obvious ways, that they use subtler forms of contact than the friendly shoulder-punch style. Or it may be that they nurse grudges a whole lot better. If your life, your children's lives, and your family's survival depends on trusting relationships, you take those relationships seriously. In that context, it may be natural for females to stay angry longer and to be less forgiving of transgressions than males, who care more for perfect strategy than perfect relationships.

As de Waal also points out, the chimpanzee males' preoccupation with status and hierarchy is not unlike that of human males. Men love to build hierarchies, too. Any traditionally male-dominated organization—from the military to the corporation—is based on a pecking order, a chain of command. Rules accommodate competition, even confrontation. "Picking a fight can actually be a way for men to relate to one another, check each other out, and take a first step toward friendship," de Waal notes. And this kind of bonding is alien to most women, who see confrontation as causing painful, hard-to-mend rifts in their feminine fabric of connections.

A colleague of de Waal's at Emory University's Yerkes Regional Primate Research Center, psychologist Kim Wallen, tells this story about a friend of his, a woman lawyer who gave up trial law: "She could never adjust to the contest nature that males brought to it. She said, 'They would say the worst things about each other in court and then go pee together and laugh at each other's jokes and I would be seething and taking it all personally.' "

Wallen has studied the way testosterone seems, in males, to spike up in competitive situations, climbing in the face of challenge, relaxing as the need to face off declines. The response occurs in a wide range of species, including primates. Does this make it easier for males to play such power games? "Now I don't know whether there are hormones involved in this," Wallen says, "but I could imagine that the fluctuations males experience could make for white-hot responses which are meaningless a few moments later."

Such fluctuations, looked at from a woman's perspective, could add up to emotional inconsistency, even untrustworthiness. Among men, if you live by competition and embrace aggression, then you must also be on guard against it in others; suspicion and a certain distancing are part of any competitor's stance. In the opening of her book *You Just Don't Understand*, which explores male-female communication styles, sociolinguist Deborah Tannen tells an anecdote about herself and her husband. They had a commuter marriage at the time, each teaching at a different university in a different city. Friends and colleagues would commiserate with them, and Tannen tended to accept the sympathy. She would reinforce the support with a sigh and agree that, yes, they did have to fly a lot and it was tough. Her husband, on the other hand, did not care for condolences. In fact, he would even reject the comments with irritation. He downplayed the problem. Tannen's analysis, after doing research for her book, was that they were following established gender roles. She was tapping into the network of connections that women like to build. He was responding like a typical male: Trust no one and reveal no weakness, lest it diminish one's status.

The way we talk to each other has received a lot of attention in recent years, but we also talk *about* each other—a lot. Gossip, stemming from a fascination with the business of others, seems to be a trait also deeply woven into our evolutionary history. Stanford University primatologist Robert Sapolsky points out that it's not just us; gossip seems fundamental across primate species. He describes, for instance, a vicious fight between two male baboons in the African savannah, going at each other with absolute fury and wicked canine teeth. A bystander, Sapolsky notes, could get hurt watching. What did the other baboons do? "Drop everything, stand up on two feet, and push in for a closer view."

Among researchers, gossip is being reborn. They are moving its reputation from sly bitchiness to an essential part of social connectedness. A 1996 issue of *Psychology Today*, for instance, declared firmly that gossip is good—a way of navigating the social environment and figuring out what's important. Good people gossip, in other words. I find myself, in a perverse way, wondering if I really want to let science turn every bad habit into a good one. Besides, what's the fun of trashing people behind their backs if it only makes you a better person?

However, the new, improved gossip research also offers insight into gender differences. Everyone gossips—boy and girl, man and woman. Generally, gossip gets intense in late adolescence. And generally it starts with talking about others of the same sex. That's right; in spite of the stereotype that teenage girls do nothing but talk about teenage boys, they actually spend three times as much time talking about other girls. Girls do, however, tell their girlfriends about boys they like—all about how great he looked in that blue tank top, and how he teased the teacher in homeroom and, well, everything.

Boys tend not to do the equivalent. Detailed research finds that teenage boys spend very little time describing the way a girl smiles or the sound of her laugh—even to their best friends. Girls do it casually; it's fun. For boys, it's too intimate. And by the time these boys and girls reach college—as documented in a study in which researchers secretly tape-recorded students in a university cafeteria (remember this next time you want to discuss your sex life over lunch)—the gender gap had deepened. College women mostly talked about people in their lives—classmates, friends, guys they dated, family. Men mostly talked sports, politics, tests, and classwork.

Books such as Tannen's, exploring the puzzle of communication differences, reveal how amazingly closely we seem to follow the ancient ground rules that appear in chimpanzee society. In childhood, the idea of status—being alpha male, if you will—seems to occur very young in boys. It doesn't seem to, at least not so much, in girls. Overall, girls seem less determined to be number one. (Of course there are exceptions.) A typical study of children's play illustrates the point by exploring the way groups of same-sex children played the game of "doctor." The boys all wanted to be the doctor—the one in charge who told everyone else what to do. They would argue this out for an extended period. Girls would ask who wanted to be doctor. Then they'd negotiate, sharing roles as doctor, or sometimes several doctors, nurse, and patient.

Boys, more than girls, want competitive games with real winners. They also seem more comfortable with confrontation over the outcome of the games. In one of the classic studies of child play, by Janet Lever in 1972, girls' games turned out to be measurably shorter than boys'. Lever saw boys arguing all the time over their games—who did what, who had what. Not once, though, did she see a boys' game end due to an argument. With girls, though, the children themselves complained that quarrels regularly ended their games. At least one girl routinely marched away, saying she didn't want to play anymore and, often, that she wasn't speaking to her fellow playmates either. It's worth raising here the parallel to reconciliation patterns in chimpanzees. Once again, males let go of disagreements more easily than girls; girls have a stronger emotional response to conflict. Some argue, of course, that this is a result of culture, that we encourage girls to be more emotional and discourage emotionalism in boys. In this argument, everything is culture—similar behaviors in closely related species are just a coincidence.

But much as we try to separate them, biology and culture aren't mutually exclusive. They can't be. Our questions about the role of nature versus nurture form a circle in which one influence feeds the other and around it comes again. We're not chimpanzees, certainly, but chimpanzees' own remarkably complex society may provide perspective on what ours might once have been. In this loop, biology is most important as a starting place. But how do you figure out where a circle begins?

Michael Milburn, a psychology professor at the University of Massachusetts, may have traced one starting place through a backwards-looking study of adult political attitudes, both liberal and conservative. Milburn was interested in whether there was a connection between growing up in a very strict household—in being severely punished for infractions—and one's adult political orientation. He found a curious and, I think, enlightening gender difference.

In general, he found that children of parents who went lightly on the punishment, especially corporal punishment, grew up to be political moderates. But children of high-punishment parents—who disciplined particularly with the belt and hand—tended to split by gender. Men, notably those who never sought counseling or therapy, grew up very politically conservative. Most were strong on capital punishment and big believers in military force. Women, on the other hand, who were whipped often as children almost always became politically liberal, whether they went through therapy or not. Milburn saw that different response as beginning with female empathy. The girls had been severely punished themselves. They tended to put themselves in the place of another person facing punishment. He argues that the difference begins in biology, in that inclination of females toward empathy, citing Hoffman's work with baby girls. Family and culture, Milburn suggests, only expand and build on what we bring into the world, which then affects our social choices, and so on.

Some male-female differences in emotional response are so distinct that it's easy to understand why scientists would look backwards for an explanation.

Ruben and Raquel Gur recently took such a route in their studies of emotion. They were following a lengthy parade of experiments which show that women are far more adept than men at interpreting facial expression. The evidence is so strong in this direction that researchers are beginning to take it as a given. So now they've started exploring why. Why are women better than men? Why should they be? One of the theories goes directly to primate societies. That is, in most one-to-one relationships in monkeys and apes, there's a dominant-submissive pattern. Put two rhesus macaques in a cage and they'll negotiate that relationship; forever more, the submissive partner will back away, lower the body, and accommodate the dominant partner.

There's always a slight element of fear in this, of the risk of annoying the dominant partner and getting hurt. A worried monkey is known to produce a kind of nervous smile, known as a fear grin. Similarly, women tend to smile far more often than men; of course, men often encourage women to smile. Researchers have pondered whether this pattern—women's ready smiling, men urging them to do so—comes from a past based on a dominant-submissive relationship. Similar questions have been raised about women's apparently sharper ability to read nonverbal expression, from body language to the subtle set of an unsmiling mouth. Is the latter from sadness, worry, or a gathering anger? Is it important in terms of self-protection to know?

The Gurs added all kinds of interesting complexity to that question. They looked at not only how well each sex can read facial expression, but whether the sex of the face studied made a difference. Can men read women's faces better or vice versa? They also wondered if there were particular emotions that drew a stronger response, say, joy versus grief.

In a series of tests studying photographs of faces, both sexes were equally adept at noticing when someone else was happy. Women also easily read sadness in a person's face, whether male or female, at about 90 percent accuracy. Men, on the other hand, more accurately read unhappiness in another man's face rather than in women's faces. They were right about other men about 90 percent of the time, but when they looked at women's faces, they were right only about 70 percent of the time. And the woman's face had to be dramatically down. "A woman's face had to be really sad for men to see it," Ruben Gur says. "The subtle expressions went right by them." His suggestion is that this may indeed spring from that ancient power balance. In a society—not so different from chimpanzees—where males hung out together and negotiated for power, it was vital for men to be able to read male faces; they might have had far less need to be tuned to the expressions of a female face. But for women, it would have been different.

The Gurs also did a study in which they asked people, both male and female, to look at a series of expressive faces. They found something that we all suspect: mood is catching. Both men and women, exposed to enough smiling faces, felt cheerful. Seeing an unrelieved series of grief portraits made everyone depressed. You could argue here that a male's tendency to be less sensitive to female unhappiness could serve him well, especially in a relationship in which he might actually induce unhappiness.

Let's make one more parallel with chimpanzees here. Male chimpanzees are bigger and stronger than females, and are known for occasionally losing their tempers and taking it out on the nearest fellow primate. Every so often, female chimps really get slapped around. It may occur to you that it's not an enormous reach, even today, to compare this with some aspects of human behavior. And in that dangerous-male scenario, it might make a great deal of sense for females to be very good at reading the subtlest of male expressions. It might mean survival. "Given the power differential," Gur says, "the higher sensitivity of women to male emotions may have evolutionary adaptive significance."

At this point, I must confess that all this discussion of submissiveness in women makes me wince. There's such a nasty little implication of women being the species' biological doormat. So, let's put this in a little more perspective. The point of all this evolutionary backtracking is not to label either sex strong or weak; it's to try to figure out where certain behaviors might have begun. With emotional sensitivity, perhaps women did gain extra abilities from their need to nurture children, to build strong support systems, and to accommodate the sometimes dangerous moods of men. Perhaps males spent more time playing power games. These are only theories. But even if we accept these theories, they don't diminish one end result, which is that today, women seem to have developed a powerful emotional advantage over men.

There are people who argue that women have learned to use their emotional and verbal skills as effective weapons. Some researchers say that mothers teach their daughters how to use words well and how to share feelings far better than they teach their sons. Studies show that mothers tend to talk to girls more about emotions and to make up stories for their daughters that deal with resolving emotional conflict. They encourage daughters more, so scientists say, to talk about their feelings. The one emotion that mothers do routinely discuss with their sons is anger, and, in that case, it tends to be about controlling behavior—as in, Don't push the smaller children off the climbing structure at school, even if they get in your way.

According to a plurality of studies, by the time children reach about age 13 or so, boys are quick to respond physically, while girls fight with words—or, often just as deftly, by withholding them and shutting someone out. Daniel Goleman, a psychologist who covers behavioral science for *The New York Times*, describes these differences as part of a continuum: First, girls learn to use emotional response to control their teenage relationships; then, women use that well-honed ability to out-manipulate men in their adult relationships. Females, he concludes, use language with more sophistication.

In this context, you notice, sophisticated doesn't mean nice. Words are sharp-edged and women handle them with practiced cruelty. If you believe the argument, men make easy targets.

There's another new school of thought, also suggesting that emotional strengths are real ones: they help make women physically healthier than men. This represents a sea change in Western societies, where emotion used to be regarded as a weak female quality; where reserve and self-control were both more male and more admired. Now there's growing evidence that weaving emotional bonds can be compared to weaving a safety net. There's a return on emotional giving. A man's traditional reluctance to reach out very far, or to very many, results in a net more fragile, more liable to betray him. Women, in general, cushion themselves with the company of others in a way that eludes men, sometimes with catastrophic results. Who, after all, are more often the loners in society—especially those at the alienated fringe? Federal prosecutors portray Ted Kaczynski, of Unabomber fame, as spending years planning meticulously impersonal murders of strangers while he shut himself away in his Montana cabin.

There are a multitude of studies that compare the reactions of men and women who lose a spouse to death. They show that women recover their emotional balance faster. If children lose a parent, it's the daughters, in general, who seem to take less of a mental beating, and who fall less often into antisocial or destructive behaviors.

But it can't be all attributed to connectedness or willingness to talk and share. There's a female something—still mysterious, even to scientists—that seems to provide the foundation for a different kind of emotional security. Frans de Waal cautions moderation, noting that we seem to be in the middle of a pendulum swing that admires female qualities, where once we venerated the male: "Women stand around in front of baby clothes oohing and aahing and everyone thinks that's just charming. Why can't men be more like that? But let a man follow his natural inclinations—maybe it's looking at pictures of undressed women—and he's really a jerk. Why can't he behave better? It's almost more politically correct these days to be a woman."

The Gurs are also cautious; at least, they warn that it's far too early to say that women stand ahead of men in dealing with emotional issues. "Obviously, if you have a measure of performance on something, and one group scores higher, we can say that they do 'better,' " said Ruben Gur. "We believe nonetheless that there is yet a lot more to learn about sex differences in the several aspects of emotional processing before we can say that women perform better as a rule."

Given those caveats, let's examine the argument that females gain an advantage from emotional bonding. A good starting point might be very early in a human's life span. In the spring of 1996, scientists reported the first results of a planned five-year study comparing the development of children in day care with children staying primarily at home; the study, sponsored by the National Institute of Child and Human Development (NICHD), is meant to be the most comprehensive look yet at the consequences of early childhood separation of parents and child. This first report followed children from the age of one month—about the earliest time that most babies go into day care—until they reached 15 months. In subsequent stages, the study will examine impact on behaviors, even intelligence, but the first focus was, basically, on bonding.

How strong is the bond between parent and child? Do daytime hours spent apart on a regular basis stretch that bond until it weakens or even breaks? Researchers who do this work try to measure this by looking for signs of either secure or insecure attachment. Secure attachment essentially means a sturdy bond; just the presence of the parent makes the child happy. If the baby is upset, he or she finds comfort in the mother's arms. Insecurely attached children are, in a sense, without that harbor. They don't always run joyfully to their parents. When they see them, sometimes they look away. And when they cry, if they do run to their mother, they may find no reassurance in her arms. Often, they push away in frustration.

The overwhelming conclusion of the child care study was that the breaker of parental bonds, it seemed, was not happy hours in day care as much as unhappy hours at home. If a mother was engaged—a giver, a toucher, a listener, a crooner—a mother expressive in her love, the bond stayed strong whether the child was at home or at day care. Those results were hailed nationwide by working mothers. The researchers were, in fact, upbraided by angry stay-at-home mothers, who accused them of diminishing their decision to leave work and be with their children.

This is not the definitive word on day care, remember. These are early results. There's no promise that the later work won't find less reassuring differences. Everyone involved with the study adds that caution. Yet one of the study's lead investigators, Alison Clarke-Stewart of the University of California, Riverside, also points out that there's a certain egocentrism in a mother's idea that she and only she can be good for her child. If we reach back into evolution, if we look at other primate societies, there's a great deal of shared child care. Monkeys rely on the help of informal babysitters. Before this part of the twentieth century, when human families stayed close together, many children were raised by a family network of parents, grandparents, cousins, and so on.

We've attempted to institutionalize that with child care centers. It's not comparable, exactly. An extended family surrounded a child with enduring relationships. A center, where people are paid to look after the children of unrelated others and where caregivers often leave for other opportunities, offers far less permanence. But in good day care centers there's a great deal of affection and touch and music and even love. As a working mother myself, I know I'm not always equal to the day, or equal to my children's demands. I may be exhausted, maybe even exasperated. On those days, I believe that my children are going to have a happier experience with someone else, a carefully chosen someone else. Do I believe that about every day of day care? Do I confidently stride away, happy in my choice, on one of those days that my schedule demands I close the child center door and leave my two-year-old sobbing, not ready to say goodbye, in the arms of a cooing surrogate? My capacity for fooling myself doesn't stretch that far.

Another of the study's researchers, Jay Belsky of the Pennsylvania State University, provides some comfort for those of us who agonize at the daily parting. It comes out of the research itself. The mothers who assert that their children are generally better off in day care—who are relaxed, casual, and relieved about leaving them there—are much more likely to have children who lack a secure attachment. In a sense, these mothers haven't pulled tight the emotional threads; they don't grieve at the door because they can let go so easily. And their children are, perhaps, more emotionally adrift. "So, guilt isn't always bad," Belsky says. "In this case, it means you care, and it means you're probably going to put some energy into making sure it's good day care as well."

Some other numbers from within the study's overall numbers are a little less reassuring and reveal another gender difference, this one a surprise: Among baby boys whose mothers worked more than 30 hours a week, 58 percent were judged securely attached. If their mothers stayed home, the warm attachment climbed to 65 percent. It's obviously a slight difference. But as noted by Clarke-Stewart, "The effect of whether you work or not is relatively small. It's how the mother behaves with the child." The striking thing, though, was that numbers moved in the opposite direction for girls. Sixty-six percent of the daughters of working mothers showed secure attachment, compared to 58 percent of those whose mothers stayed home. "We never expected that," Belsky says. The expectation was that the attachment would also be higher for small girls at home.

On the other hand, he adds, there's nothing surprising about the suggestion that boys are more emotionally vulnerable. He found the same pattern himself, in a smaller 1988 study which found that too much time in preschool seemed to weaken the bond between boys and fathers as well as mothers. Says Belsky, "There's plenty of evidence over the whole life that males have a more vulnerable biology. They succumb to all sorts of things faster than females. In that context, it might make sense that they would be more vulnerable to the stress of a separation. Maybe it's just more difficult for boys to feel secure." It's something you might expect from an evolutionary history of regarding relationships with caution.

"The boy finding takes an added significance because it's consistent with other data on male vulnerability," Belsky says. And research suggests that baby boys need more of everything from their mothers; for example, they tend to be born slightly later in gestation, and they're bigger, requiring more feeding. From the beginning, the basic investment is greater. Still, the finding that baby girls were, overall, slightly more secure if they spent time away from home was so unexpected that the researchers with the NICHD study debated whether they were just misreading the data.

"We're all at a loss," Belsky continues. "It's definitely a finding in search of an explanation." Perhaps it's only a random glitch. Or perhaps not. There are suggestions that girls whose mothers work develop into more achievement-oriented adults; their mothers become role models.

"Is it possible that they're picking up something along those lines in infancy?" Belsky asks. Or, he suggests, it could be that the emotional bond between mothers and daughters can get too tight, too close, enmeshing rather than securing, so that the daughters who stay home actually may be made more fearful. Harvard psychologist Jerome Kagan, the leading expert in the biology of shyness, has shown that if parents give a shy child some extra room —if the child is allowed to make mistakes and allowed to bounce back—the result is a gain in confidence and a reduction in fear. It's the shy child who is overprotected, bundled, and muffled against the world, who stays fearful of life outside the house.

"Maybe young girls need an opportunity to explore and feel out new environments," Belsky says. "And maybe this study will get our noses pointed in the right directions. It's possible that we're pushing boys toward more independence than they can manage, and that we're giving girls more cuddling than is in their best interests. We're exaggerating a proclivity too far. Maybe this study will help us get a better handle on the meaning of both [the needs of boys and girls]."

The study's next stage assesses children at ages three to four. Belsky suspects it may offer a better indicator of what's going on with boys. That is, if they turn out defiant and aggressive, that will be pretty visible. Troubled girls tend to be unusually quiet at that age, a little more fearful and withdrawn, and that's less visible and harder to measure. As Kagan points out, emotional connection is an elusive thing to measure at best, and how we build it undoubtedly starts well before a mother carries her child into a day care center.

Biologists like to compare the brain of a newborn human to that of a fetus—unfinished at birth, the most incompetent among all primate newborns on Earth. They also have a good explanation for why that is: it has to do with the size of the baby's head and the width of the mother's birth canal and pelvis.

Let's put it like this. If, in fact, our species needs a female who can walk on two legs, then the basic pelvic design is about as big as it can get. That means that the average size of a baby's head can't get any bigger. But if we also want to be a smart species—and we certainly claim to be—then we can't be held back by the limited brain size the birth canal is able to accommodate. The compromise is a very unfinished brain at birth. The newborn human possesses a brain only one-fourth the size of an adult's; by comparison, a monkey's brain is almost two-thirds finished at birth.

Any other animal—any other primate—would be considered defective if it had only the level of competence that a human possesses just after birth. By the time a human infant is just starting to roll over, a baby chimpanzee is walking and playing with its companions.

We make up for it later, however; the brain doubles in the first year of life, then doubles again. In its own way, that open-endedness is a wonderful aspect of humanity. It does make us remarkably floppy at birth, but also flexible beyond any other species—born with less in place, and therefore more open to experience. And it means that what happens to us—right from the moment our parents hold us tight or, in a dark reverse, slap us down—is enormously influential.

Because our brains are so immature at first—because we haven't even laid down all the apparatus necessary to hold memory—we may never recall whether our first touch was gentle or cruel. We are left not completely knowing what bonds were built and broken in those first days, even years, of life. Do those encounters, the ones we won't remember, shape our brains in ways we still don't understand? Marc Breedlove says there's every reason to suspect that many experiences, especially emotionally powerful ones, physically alter the brain. "Let's take an obvious example, the assassination of President Kennedy. It's become almost a cliché, but if you ask most [middle-aged and older] adults today if they remember when it happened, they remember the event, where they were, and even whom they were talking to. And for that to happen, it had to have been an event that altered their brain so that it stored the information. We don't remember it because we hold the memory somewhere in our soul—we remember it because it changed our brain."

We don't have to remember the experience for it to be mind-altering, or for it to be permanently written in the rules of our behavior. All kinds of things happen in those first days of life, when the brain is still too basic even to have good capacity for memory storage, and yet as adults we continue to apply those early lessons. Does anyone remember learning to drink from a cup, or being taught the right amount of gentleness with which to stroke a sleepy cat? Our brains acquire the skills, even though they cannot recall the lessons. Breedlove's favorite example is language. Who remembers being taught to speak? What adult holds a recollection of parents reading board books over and over, repeating simple words almost endlessly ("Say goodbye, honey, say goodbye"); and patiently naming the world—cat, tree, table, sky. And yet that experience, in those early months, is essential to development of language; without it the brain never completely puts the words into their proper place. "We don't remember the experience of learning language," Breedlove says. "But that doesn't mean we didn't have the experience and that it didn't change us and our brains. Of course it did."

There's a clear indication that emotional learning and connection is also enormously important at a very young age. In a brain study of very young children by University of Seattle psychologist Geraldine Dawson, she found that lack of affection produced physical results. Dawson did PET scans of the brains of children whose mothers were severely depressed. They showed, she reported in 1996, that metabolism levels dropped severely in the brains of those babies compared to those with warm and cheerful mothers. Dawson particularly saw a difference in the frontal lobes, where the brain seems to handle positive emotions like curiosity and happiness. The normal building of connections in that region did not show up in the children with depressed mothers. Dawson also found that most of those children began showing signs of angry and aggressive behavior by the age of three. The study suggests, at least circumstantially, that the unhappy infancy produced a brain less open to happiness and more prone to negative actions.

What Belsky wonders, then, is whether somewhere in this tale of emotion and health and developing brain is another connection that we've somehow missed or misunderstood. Perhaps, especially early on, boys need more—more touch, song, emotional support —than we tend to give them. Maybe parents disconnect from boys early in a way they don't with girls, pushing them faster toward independence. As grown men, they wouldn't remember that, of course, but the emotional distance may have been built in anyway.

You have to wonder if our culture has misread the realities of our biology. Remember that old nursery rhyme, that girls are made of "sugar, spice and everything nice" while boys are "snips and snails and puppy dog tails"? I quoted it carelessly to my son Marcus one morning when he was five. He was outraged. Why did girls get to be the good stuff? Why did boys get stuck being garden pests and, apparently, amputated dog body parts? Maybe we need to accept that, at least very early on, baby girls may be stronger than we know and baby boys more vulnerable. In this case, stereotypes may do neither sex any favors.

The late Nancy Bayley of the University of California, Berkeley, whose tests of infant development are still standard, once compared the effect on infant boys and girls of mothers who were depressed enough to be withdrawn or hostile. These were long, awful depressions in mothers often resentful of carrying the extra burden of a child.

Bayley used the first 18 months of life for her comparison, charting children who received affection against those who did not. She then waited until the children were a little older, between the ages of three and five, and returned to do a series of intelligence and competence tests. Her results suggested that the boys raised by withdrawn and hostile mothers might indeed have been harmed. Their intelligence scores were steadily below those of boys raised with tangible love. That wasn't the case with the girls. They were more withdrawn than children from happy families, but there was no apparent effect on intelligence. For girls, no matter what the home life, the test scores correlated best with the tested intelligence and social class of their mothers. There was one exception: Being raised by a restrictive mother—one who narrowed their experience and sheltered out the world—brought their test scores down. Laura Allen, the UCLA neuroscientist who has done so many comparisons of gender differences in the brain, once analyzed the Bayley work like this: "I think boys need more one-on-one attention. I think the affection may change the sex hormone level in the brain, which then affects brain development."

Why would our biology be so codependent, so that failure of affection would alter the way a baby grows? Perhaps this began simply indeed, in the life-sustaining bond between mother and child. Our species has evolved as intensely social, woven together in a fabric that tears along emotional lines. There's some remarkable work by Sally Mendoza, a psychologist at the University of California, Davis, that suggests that even distant relationships, as they shift, can change the hormone balance within. If a coworker changes jobs, leaving a new opening at the office, and the familiar balance of relationships changes, our stress hormones rise. Essentially, we seek stability in our relationships with others. There's comfort in it, and safety; breaking bonds predicts only trouble. On a much larger scale, Mendoza suggests that some of the rising crime and cruelty of the late twentieth century may have something to do with our increasingly impersonal society—that we aren't biologically equipped to function in a world that demands distance.

Neuroscientist Bruce McEwen, of Rockefeller University, has shown that the hippocampus—that home for recollection, mapmaking, and much that we probably don't yet know—responds almost directly to stress, with cells dying in response to rising levels of cortisol. In studies using rats, this process of cell loss is increased dramatically if the animals are isolated and left without social support. In humans, there are countless studies, of countless men and women, showing that they best survive even severe illnesses, such as heart disease and cancer, if there's a loving and supportive marriage. Even a minor marital spat sends shivers through the immune system, so that couples after a fight are much more likely to come down with a cold or the flu. Scientists have found that, if a partner dies, the immune system of the survivor stays depressed for between four and fourteen months. And no one has ever found that loneliness or grief are good for health, mental or physical.

We really don't understand, however, why boys and men seem even more vulnerable than girls and women in these respects. Is this, too, a consequence of our times? Belsky, Kagan, and others have suggested that emotional isolation and its negative results are simply part of the male package—less durable than the female design, linked to a live-fast-and-die evolutionary past. But now, as medicine keeps us all alive longer, and as technology and changing culture make us more independent from one another and less bonded to extended families and communities, the difference becomes more pronounced. And it becomes more destructive.

Yet even if all of that is true, it may merely add up to a classic example of how difficult it can be to apply scientific theory. It's easy to say boys would be better off with more cozy, chatty relationships, the kind that girls tend to establish and maintain naturally. But how do we make that happen across a society trained in a clearly opposite tradition? Further, as the Gurs reminded me, we're early into this science; we're still sorting emotional strength from emotional weakness, deciding what the terms mean.

Even as sexual science can't nail down every intangible nuance of poetic romance—not yet, anyway—there are surely limits to what research into emotions can tell us. Let's dwell, for a moment, not on love but on another powerful emotion: grief. Suppose we do probe into the architecture of the brain and discover in women a different processing of grief than that in men. In an overall sense, yes, that would be fascinating. But at the individual level, could you take that to a couple who had lost a child? You might use it to explain why the husband and wife handled loss differently, why perhaps he withdraws and she sobs on everyone's shoulder. But you can't predict individual behavior from these studies; it's the overall portrait that emerges. You can't argue that every woman is a good emotional communicator, or that every man is a stiff. And you can't argue, ever, that only women love their children, or suffer in their loss. There's a point at which science cannot—and probably should not—define all aspects of who we are.

I've read elegant neo-Darwinian arguments concerning why we grieve for lost children. One of the most persuasive was made by Robert Wright, in his study of evolutionary psychology, *The Moral Animal.* Wright dissected Darwin's life as an example of how genes may influence our behavior. And he offers them as a partial explanation for Darwin's enduring grief over the death of his ten-year-old daughter, Annie. After his child died of a lengthy illness, Darwin wrote in his diary, "She must have known how we loved her. Oh, that she could now know, how deeply, how tenderly, we still do and shall ever love her dear, joyous face." We can read into that how alike men and women are, how all of us share that fear of not communicating our emotions adequately, of losing people we love without being sure that they knew how much.

Now Wright suggests—and he admits in the most charming way that this is a cynical analysis—that part of Darwin's grief is based in a kind of silent genetic calculation of reproductive loss. That is, Annie was ten; she was nearing the time when she could pass on Darwin's genes. The loss is much greater than if she were, say, forty. He also cites evidence that if people are surveyed about losing children, their grief is most intense when there's a reproductive potential lost; sorrow reportedly declines as the child ages into infertility.

By this standard, grief over the death of a child becomes a sort of internal loss inventory—"Oh damn, forty-six perfectly good human chromosomes wasted." And this is where I start parting ways with a totalitarian view of evolutionary psychology. Maybe all our behaviors can be dismantled and reduced to the click of the genetic calculator. We're a long, hard way from proving that. And I doubt we are that limited. Further, if you take loss and grief down to this chilly level of analysis, do you really understand them more anyway? If we put men and women too tightly into different emotional boxes, we run the risk of understanding each other less, rather than more. So I want to end this discussion in a less scientific way. Let's get emotional. Twenty years after his daughter's death, Darwin was still mourning her loss, still confiding to his diary how much he missed her. We can match that with another tale, equally sorrowful—this time from the mother's point of view.

Not so long ago, a friend of mine at the office was worrying about her sister in Minneapolis. A year or so earlier, the sister's eight-year-old daughter had been killed. The child's father was driving her home from a piano lesson when a speeding truck, driven by a drunk, slid across the road, smashing into the passenger side of their car where their daughter sat. The dad was only slightly injured, but the little girl was dead before the ambulance came screaming up to help them. They buried her in a city cemetery.

Now, long months later, the husband was being transferred to an office in another city. My friend's sister was planning to have her daughter's body dug up, moved, and reburied. She just couldn't leave the child there, alone, with no one to brush the Minnesota snow from her grave. She wanted—needed—to be able to stand by the grave and talk to her daughter.

I remember my friend and me—standing together by a window at the newspaper building, with California sunlight coming through, luminous as in no other place I've lived—so self-righteous, tsk-tsking to each other about her sister as we stood in that bracing spread of light. Life has to go on, she should leave cleanly, start anew, get on with it. After all, her husband wasn't doing this neurotic number.

And yet it kept coming back to me, the little grave, abandoned in the winter. Until I realized, slowly and so reluctantly—I did not want to admit this—that I too might be just that lunatic; that I too would find it impossibly hard, if I moved, to leave a child of mine buried here. Because, you see, it would be like leaving behind the point of my heart.

Chapter Four

PERFECT PARTNERS

The Monogamy Puzzle

Monogamy is a mystery story. If you must have two sexes, then why complicate things with, well, relationships? And then why make some of those relationships inconvenient—not expeditiously transient, but lifelong? Why would such a profound commitment come about in a biological world which allows sexual relations to be as noncommittal as the lazy drift of a pollen-dusted bee between two flowers?

The planet's monogamous species are rare because promiscuity is easier—and it works. With monogamy, we're looking at an exception to a life rule. Birds are the one group in which it's predominant. Among them, an estimated 90 percent of species are apparently monogamous. But in mammals, only an estimated three percent of species are considered monogamous. With primates, it's perhaps 12 percent—if you include borderline cases such as humans.

And as any married person—or any person in a long-term partnership—knows, commitment is a demanding choice. In fact, humans aren't as dedicated to fidelity as some other species, even including some species of mice, which choose a mate and are genuinely separated only by death. Human culture worldwide is built around marriage and committed pairs, but the line in the Western marriage ceremony " 'til death do us part" is hardly a given for all.

With rings and garters and vows about "forsaking all others," we tend to think of monogamy in terms of pure sexual fidelity. A "good" woman is a one-man woman. A "good" man doesn't cheat on his wife. That moral focus used to permeate even science. Biologists would identify a monogamous species and expect mutual sexual dedication. It turned out that things were more complicated than that. (Aren't they always?) Scientists found that if they demanded complete sexual fidelity—no cheating, ever—as the definition of a monogamous species, they would rule monogamy out of existence.

While it's true that greatly reduced promiscuity is the first definition of a monogamous species, it is equally true that many cheat when the opportunity arises. Patty Gowaty of the University of Georgia has done some enlightening work on this with bluebirds. These charming, bright-colored birds tend to mate for life. They also like to enhance their opportunities—if a mate is off feeding, the spouse may well sneak off with a neighbor bluebird. These days, almost any journal on animal behavior will feature at least one article on "extra-pair copulations," meaning the discovery of sexual cheating in yet another monogamous species. Thomas Insel, a neurobiologist and director of the Yerkes Regional Primate Research Center in Atlanta, studies prairie voles, a species of fluffy rodent known for dedicated monogamy. Vole couples raise all their children together and, on a cool evening, can be found sitting side by side, the picture of a cuddly couple. But, Insel says, that doesn't mean they don't sleep around. "What we say about prairie voles is that they'll sleep with anyone but they'll only sit by their partners," he says.

I'm not arguing this as some biological endorsement of slipping a few one-night stands past your spouse. What I am arguing is that we miss the point by being obsessed with sex when we consider monogamy. In a world dominated by polygamy—a world, remember, in which 97 percent of mammals live by the rules of promiscuity—it doesn't seem surprising that total fidelity is elusive. After all, reproductive success is about what works—and that obviously will vary depending on each species' situation. We should expect some flexibility.

The aspect of monogamy that most fascinates me is that which encompasses partnership—an equality in housekeeping, child rearing, and other aspects of a relationship. Monogamy, the kind that the prairie voles practice, promises a male-female sharing of chores that you never see in a polygamous species. This is not to say that all monogamous species share equally in life's work. There's incredible variation among monogamous species—from the sexually faithful males who do little child rearing (such as African dik-diks) to the equally loyal male partners (such as South American titi monkeys) who do most of the child care. But as a rule, only monogamous species build solid, and even equal, male-female partnerships.

So as a human female—often driven to wonder why equality is so hard to realize—I look at biological monogamy as a way to gain insights into the so-called traditional roles of men and women. Many anthropologists think we evolved from largely polygamous hominid species. Does that predict a role division in our society? Does it explain how we might pick a partner, what attracts us to each other? Some biologists say that physical appearance provides many clues as to relationship, including whether a species is monogamous or polygamous. And so I wonder—what do *we* look like on that spectrum of behavior? Do the shapes of our bodies hint at promiscuity or fidelity, partnership or separate responsibilities?

Biology proposes that the details of the way many species look—shape of face, length of leg, curve of hip—are grounded in competing for the one-night stand. That's the theory of sexual selection, a basic in the understanding of polygamy. If you're an animal that is not interested in finding someone who will be a compatible, caring partner—if all you want is a good, healthy egg for your healthy sperm, or vice versa—how do you choose? How does the female decide which male will provide the best stock for strong children? What tells the male that he's looking at the ideal vessel to carry his genes on into immortality?

In this polygamous scenario, advertising is everything. Appearance is nature's version of a personal ad: single male, great muscles, spectacular build, flashy mane/tail/teeth. In other words, good genes. There's evidence across a range of species that some of the most spectacular, and even beautiful, shapes and colors have to do with the show-and-tell nature of mate competition. Charles Darwin himself proposed sexual selection more than 100 years ago, based on careful observation and his own belief that everything in life has a reason behind it: "It cannot be supposed that male birds of paradise or peacocks, for instance, should take so much pain in erecting, spreading and vibrating their beautiful plumes before the females for no purpose."

That's so taken for granted these days that we often forget how Darwin's theory flew in the face of one of the prevailing religious beliefs of his day. To the church, the vivid beauty of the peacock was another testament to the design talents of the Creator, who had fashioned for us—his most important creation—a lovely setting in which to live. Darwin's major sin, of course, was undermining the Victorian Christian view of the world. And part of that was his pulling humans into the natural world, refusing to separate us from other living species. Darwin, in fact, made a short, sharp leap from peacock to human: "A girl sees a handsome man and, without observing whether his nose or whiskers are a tenth of an inch longer or shorter than some other man, admires his appearance and says she will marry him. So, I suppose, with the peahen."

From a late-twentieth-century viewpoint, of course, his description of women makes them sound pretty dim. But at least he grants them a say in the matter; the theory of sexual selection as shaped by Darwin's successors seemed to ignore female influence almost entirely. Biologists argued that everything interesting about male body shape came from male-male competition. It was still related to the one-night stand, but it was about the males duking it out for the sexual prize. The female in this context apparently stood by like a lamppost and then meekly accepted whoever won, even if he did so accidentally—say, through his opponent's falling over a tree root during the fight.

To be fair, there's some truth in this view of mate selection.

Male animals of many kinds really do battle for sexual dominance, and there are some aspects of appearance—the neck-protecting lion's mane and the weaponry of the aptly named bighorn sheep—which seem to have evolved to equip them for such fights. Even Darwin believed male physical rivalry was much more important than female attraction. Despite his fascination with the peacock's tail, he considered such decorative displays to be secondary. But today, science has come to realize that female choice is at least as influential as male dominance, if not more so.

Like the flash of a peacock's tail, the red crest of a rooster's comb (apparently a real turn-on to hens) is an indicator of good genetic inheritance. The hen doesn't choose a rooster in quite the way a particularly calculating woman might choose an all-genius sperm bank, but their goals are certainly similar. And just as the sperm bank makes a certain claim of reproductive advantage, so does the cock, through his comb.

That upright, brilliant-colored ornament is an advertisement of high-quality roosterhood. It stands for what the hen wants, and, not coincidentally, it gets the rooster what he wants. Over generations, hens' preference would encourage comb development in roosters, as those with the best headgear were most successful in getting mates and in passing on their genes. The hens' bias would also be reinforced in successive generations, as male offspring of such chickens attracted the most mates and thus passed on the genetic advantage.

In 1982, the great Swedish geneticist Malte Andersson provided a clear illustration of female-driven sexual selection among African widowbirds. Male widowbirds drag around spectacular tails which approach feathery excess, some reaching more than six feet in length. To test what the tailfeathers mean in terms of mate attraction, Andersson decided to alter some of them. He glued extra feathers on some tails, snipped some short, and left others exactly as they were. It turned out that even the showiest natural tail couldn't compare with the attraction of Andersson's creations. His augmented males cheated nature, rearranging the widowbird mating season. The newly enhanced fellows were suddenly in great demand, sharing new nests with available females. The poor sheared birds, by contrast, had a terrible time finding mates. And perhaps saddest of all, the widowbird males whose naturally long tails would under normal circumstances have made them the stars of the season found themselves displaced in the mating hierarchy. This clearly had nothing to do with male-male battles; the tails are useless as weapons, with long-tailed birds no stronger or fiercer. It was all simply about looks. Similar games have since been played with many other tails, from those of peacocks to those of barn swallows. All have yielded the same result.

It's sometimes difficult to tell how females end up with the genetic "Mr. Right." Jane Brockmann of the University of Florida has been investigating that question with those strange, prehistoric-looking creatures called horseshoe crabs. Dark and shiny, looking unsettlingly like spades that have crawled to life off a deck of cards, these ocean dwellers seem so self-contained in their heavy shells that they might be mistaken for something completely asexual. Laughing, Brockmann explains that the horseshoe crab is far from that. Still, the animal's alien appearance does count for something. She says the crabs look so weird that she doesn't have to bother with the political baggage that so often weighs down sex research: no one mistakes her work for commentary on relations between men and women. "It's like reporting on the sex life of Martians," she says.

And yet, within those seemingly impervious shells, horseshoe crabs behave in ways that are not so alien. The researcher has found that crab females tend to routinely end up with the stronger, more vigorous males.

Scientists can pick the studs out easily on sight, by their clean-cut look. As with humans, horseshoe crabs will lose their glossy, smooth appearance with age. Technically, the crabs' protective coating of slippery mucus wears off and they can become encrusted with barnacles and bedraggled with seaweed. Is this a big "ugh" for the females? Does the bumpy, wispy look advertise diminished virility and perhaps a less potent package of genetic stuff?

That's not clear, although Brockmann has found that when the crabs float ashore, clasped together in mating position, the females are usually locked together with the gleaming youngsters. If she was studying "female frogs or crickets or female grouse, which actually march up to and pick out males from a group," Brockmann says she'd be sure that she was observing clear female choice. But analyzing horseshoe crabs is trickier. She suspects that there may be a kind of "passive" female choice, that perhaps only the younger, stronger crabs can maintain the sexual clasp and, therefore, achieve the embrace that leads to reproduction.

The easiest analogy to physical sexual advertising in human males is height: tall men are routinely rated most attractive by women, and, in fact, are more appreciated by all of society. There's nothing subtle about the connection between height and power. Studies have shown that tall men seem to have an advantage in seeking jobs, in getting promotions, and on the pay scale. Between 1904 and 1980, the tallest candidate won the U.S. presidential election 80 percent of the time.

But there are more subtle physical signals among males as well —a neatly symmetrical face, for example, or a beautifully chiseled pair of cheekbones. It appears that all nature loves symmetry as an advertisement of genetic fitness—and asymmetry becomes the obvious opposite. Female scorpion flies prefer mates with identically sized wings. Just so with humans: babies smile more at symmetrical faces; girls and women seem to like a balanced male face. Research shows that boys with more symmetrical faces start having sex some four years earlier than their lopsided competition, and that they pump out more sperm when they ejaculate. It turns out that men with less symmetrical faces (yes, researchers take fraction-of-an-inch measurements) report more health problems. They also report more anger, jealousy, and insomnia (although it's not clear, at least to me, whether that is connected to better genes or to relationship problems). But scientists who study human concepts of beauty and sexual attractiveness point out that testosterone is key to developing some of the notable aspects of male beauty. Height, a wicked lower jaw, and masculine cheekbones are all strongly influenced by steroid hormones. There's some slim evidence that the most symmetrical men average slightly above average in circulating testosterone. The connection may hold also for women and estradiol; for instance, symmetrical women are reported by some to be more fertile.

Scientists call these shape differences secondary sexual characteristics. That means that even if they're not a direct component of the reproductive system, they belong to the mating package. They may, if nothing else, be part of the basic "come hither" needed to begin the action. But good looks don't come cheap.

Sally Mendoza, who chairs the psychology department at the University of California, Davis, offers this perspective on the pros and cons of sexual advertising: "The adornments in structure that are believed to evolve with female choice in most cases put males at an extreme disadvantage with fighting [such as long tails] or predation [bright colors]."

There is also increasing interest in the notion that these characteristics are disadvantageous physiologically. Testosterone is generally believed to inhibit immune function and most of the male secondary sexual characteristics are testosterone-based. Therefore, the animals that are most preferred by females are somewhat immunosuppressed. Why would females prefer such males? Because only those males that are in extraordinarily good shape to begin with can afford such a flagrant display.

There's ongoing debate over how—or whether—women signal back to men. Is there a comparable display of female reproductive potential? Researchers have spent years calculating the importance of waist-to-hip measurements in women, right down to the decimal points. The idea, basically, is that the perfect curve is engineered by the perfect amount of fat, which is roughly comparable to the amount of calories needed to make a healthy child (80,000). Men respond to women's fat-padded curves because they signal reproductive health. Feminist anthropologists get fairly sarcastic on this point, suggesting that the idea be renamed the "tits and ass theory of evolution." Why should everything about a woman's body have to do with male interests, they argue. Maybe women's bodies are simply well-designed for producing and raising healthy children.

It could be argued that our obsession with super-thin fashion models runs counter to the notion of sexy curves, but female models are hardly chosen for reproductive purposes. Their first job is to sell clothing to other women; stirring a response from men is their (implied) second job. On the other hand, the modeling world's obsession with youth—successful models may start at age 14 or younger these days—does fit into the evolutionary pattern of sexual advertising.

This last isn't as pedophilic as it superficially appears. Evolutionary psychologists point out that, in prehistoric times—and a lot more recently, too—if a male was trying to pick a mate who was likely to bear healthy children, he would be better off with a younger woman, perhaps very young. People lived in tight-knit groups, and it was probably very apparent that as women aged, the pregnancy success rate began to falter. That hasn't changed today; we just understand it better. As a woman's eggs age, their genetic perfection cracks a little, and the odds of chromosomal mutations increase far too fast. Take Down's syndrome, for an example. It happens when the two number 21 chromosomes in the egg fail to separate as they merge with the corresponding chromosomes in the sperm. Thus, while normally a child gets one of each chromosome from each parent, in Down's, a child inherits three copies of chromosome 21—two from the mother, one from the father. This is called a trisomy and the effect, as we all know, can be devastating. The risk of Down's goes up almost unbelievably abruptly as women get older. The odds are 1 in 1,667 for the child of a 20-year-old woman; 1 in 385 when the mother is 35; 1 in 106 at age 40.

Even knowing that, we can't fix it—not yet. We can screen for Down's early in the pregnancy, through techniques such as amniocentesis. Doctors can tell parents if their developing child in the womb is free of the defect or if they've found the trisomy, the odd extra chromosome. At that point, the best that medicine can offer parents is a horrendously difficult choice—whether or not to continue with the pregnancy.

In the early days of human society, though, there was no way to see ahead, to even make a choice. We assume that there were, instead, observably higher numbers of failed pregnancies or children born to die as the mother's age increased. (And, undoubtedly, age 30 was pretty old back then, and a woman who made it to 40 would have been an elder.) That might have encouraged males to pick younger mates or men who did so might have been more genetically successful, selecting for a tendency in males to choose youth. Further—this is evolutionary psychology in a clearly speculative mode—this was obviously before calendars and clocks. Probably, males would have used appearance to judge age: the smooth skin and full lips of youth, for instance. There's even an argument that the youth-as-health association explains many males' fondness for fair hair: There are more blond children than blond adults (their hair darkens as they grow older).

Is men's obsession with younger women—and the tendency to leave an older wife for a newer, younger one—woven into our biological history? Is this an enduring preference, a leftover of a polygamous past, of mate competition and peacocklike visible sexual suggestion? Perhaps, but the answer is not that obvious. Some scientists suggest that fascination with youth is, in fact, evidence of monogamy. After all, if you belong to a species that mates for life, there's great incentive to pick well when shopping for a mate. At its simplest, that means, for males, choose healthy and, perhaps, choose young. On the other hand, there's no obsession with youth in our closest comparative primate, the polygamous chimpanzee. A male chimpanzee is indifferent to whether he mates with a teenager or a middle-aged matron, as long as she is in a hormonal state of come-hither readiness, also known as ovulation or estrus.

But, in general, monogamy doesn't seem to support show-and-tell looks. Species that pick a life partner seem remarkably unimpressed by vulgar tailfeather displays or muscle-bound competitions. That's because what makes monogamy work is that it is work, an arrangement in which both partners contribute to a life together. Rather than the showiest or strongest opposite number, a female looks for the male that behaves the most compatibly, often something like herself. She's seeking the match with which she can share homebuilding and home defense, the rearing of young, playing and cuddling. Courtship often involves the male offering food to the female—good-provider stuff—or showing off his particular territory. What the female looks for, in many cases, is not a good-looking guy as much as an all-around good guy. In light of this, maybe that common complaint of lovelorn American men—the one about women picking good-looking jerks over ordinary nice guys—qualifies as anecdotal evidence of the human tendency toward polygamy.

If you think of monogamy in terms of equal partnership, especially in terms of raising the children, then it becomes pretty obvious why birds are teamwork stars of the planetary show. Females lay the eggs and after that either parent can keep the eggs warm, feed the hatchlings, defend the nest, help the little birds learn to fly. Mammals on the other hand—sheep, tigers, cows, wolves, us—nurse their young, a function that is not naturally shared. In biological terms, it's the female's job. And there are increasing data about what an important job it is.

In terms of sophisticated species development, researchers have nothing but wonderful things to say about lactation, nursing, and the devotion of mothers to their young. Recently, they've been almost giddy with it. Consider a 1996 conference, sponsored by the New York Academy of Sciences, dedicated to the Integrative Neurobiology of Affiliation, which is a fancy way of saying emotional bonds. By the meeting's end, the suggestion was that we'd have stayed stupid if not for breast-feeding: "The neuroendocrinology of lactation may be important to the wiring of the mammalian brain," said C. Sue Carter, of the University of Maryland. "Its development was revolutionary," added Cort A. Pederson, of the University of North Carolina at Chapel Hill. He continued: "Sustained maternal protection and nurturing of offspring until they were able to fend for themselves allowed a much higher rate of survival. Mothering also permitted a much longer period of brain development and was therefore a prerequisite for the evolution of higher intelligence. Species that mother their offspring have come to dominate every ecological niche in which they dwell."

In our species, breast milk really does separate women from men in terms of caring for infants. It's not that a man can't feed a baby. Of course he can mix up formula and fill and warm a bottle. Of course he can hold a bottle of expressed breast milk to baby's eager lips. But he can't produce that liquid. For all the liberation the bottle gives thankful Mom, and for all the help willing, gentle Dad can be in raising a healthy child, research is showing more and more that the bottle is not a comparable substitute for the breast.

Breast milk is such a remarkable assemblage of hormones, sugars, proteins, fatty acids, vitamins, minerals, and immune components that you wonder how we ever had the gall (what the Greeks used to call hubris), to think we could casually match it with a factory-made powder or liquid, poured by machine into shiny round containers, and lined up on grocery store shelves like soup cans.

Formula is food, but there's increasing evidence that human milk is both that and biological strategy as well: one of the greatest threats to human infants is the bacteria that attack through the intestines, causing a wasting diarrhea. Many of these unfriendly microorganisms have evolved to latch onto distinctly curved cells that line the intestines. It turns out that some of the sugars in breast milk have also evolved to mimic the shape of those intestinal cells. As a result, scientists think, bacteria mistakenly grab onto the sugars and are cycled right out of the body.

In the United States, formula-fed babies average twice the number of intestinal infections of breast-fed babies.

Human breast milk also contains fatty acids not found in formula in the United States (although recently added in Europe and Japan) which promote brain development, particularly in the vision system. It also contains a type of iron which is absorbed far better by the infant body than the iron used to enrich baby formula. Iron also is considered essential to the developing brain. As Israeli endocrinologist Yitzhak Koch puts it, "Milk is not just a source of nutritional elements, some simple mixture of fats, sugars, and salts. It has a causal effect on the infant. It's really a vehicle by which the mother transfers information to the child."

But breast milk also helps explain why full partnership is rare among mammals. Nursing creates a division in child care, loading disproportionate responsibility onto the female. It doesn't rule out co-parenting, but it makes it harder. Among birds, by contrast, division of labor isn't a problem.

But even birds don't always stay with the same partner. Occasionally, they seem to believe they've made a mistake. Biologists talk about divorce and remarriage among birds, just as in humans. Scientists call this serial monogamy: one partner at a time, but one partner after another. As evolutionary biologist Judy Stamps at the University of California, Davis, points out, it makes sense: Does it really serve a female bird to stick with a male partner who turns out to be a bad choice, who doesn't really help raise the young? Note that when female birds move on, their yardstick for a good partner stays the same—they're looking for a helpmate, not the forest's greatest stud.

That helps explain why species that build such dedicated partnerships tend to be what biologists call monomorphic. Even as sexual dimorphism refers to two shapes for the sexes, monomorphism tells us that male and female have a lot in common, including an extraordinarily similar body. It reflects the fact that in monogamous species, different features may be favored. Because both sexes perform a lot of the same tasks, there's no evolutionary pressure for one sex alone to have showy muscles or big teeth. There's no pressure to show off in a purely physical way. The pressure is to look like a good parent; there's no need for obvious sex appeal. And the result is that anyone can tell a polygamous bird species (peacocks) from a monogamous one (Canada geese) just by comparing the likeness of males and females.

So how might that apply to us—clearly no mirror image of each other, and clearly sexually aroused by the physical differences? Well, we don't choose partners—or even friends—purely by appearance.

We do demand (mostly) more than looks. We aren't like cardinals: scarlet red male to dust-brown female. We don't have the enormous dimorphic size disparity seen in gorillas—among whom the average male can weigh twice what the average female does. We have huge men and tiny women, certainly, but we also have little men and big women. On average, men are bigger, but not by such a huge margin. Still, men and women are shaped differently and our cultures often emphasize the relatively subtle sex differences.

In many societies, including the United States, people reverse the bird pattern of showy male, drab female. In business and formal wear, men wear simple, darker clothes. Women may choose to do so too—to make a power statement, for instance—but they also wear feature-enhancing makeup, glittery jewelry, and bright colors. A woman's power suit can show off her legs. It can, if she chooses, be bright red. A man's? Never—at least, not if he wants to be taken seriously.

We also don't behave like an indisputably monogamous primate—like the South American titi monkey, for instance. Titi monkeys are pretty much monomorphic; only a knowledgeable biologist can tell them apart. They also carry monogamy to an interesting extreme. They aren't always sexually faithful; the female, in particular, is sometimes curious about other males, and the males consequently tend to be more jealous than the females. But the partnership dominates their lives.

They take almost two hours to get to sleep at night, the male and female checking out the other's wishes and needs and wants with such thoroughness that they can only inch toward sleep. Male and female titi monkeys tend to be so wedded to each other that, in terms of affection, their offspring come a distant second.

Most of this research has been done at the California Regional Primate Research Center at the University of California at Davis, which houses the country's largest collection of titi monkeys. When the Center's psychologist Bill Mason, who put together the titi monkey group, and his colleague Sally Mendoza pulled apart titi monkey families—separating male from female, infant from parents—they found a fundamental difference in biological response. When they checked the chemistry of stress—rising cortisol levels in the blood—the parents became extremely tense when separated from each other, the cortisol climbing high. But the hormone barely rippled in the blood when the infant was removed. It wasn't as if they didn't respond to the infant; both parents would move to retrieve it. They just weren't as upset about temporarily losing the baby as they were about being apart from each other. And in this partnership, somehow, maternal patterns have shifted to a point that all our lovely stereotypes about nurturing mothers have to be abandoned. Titi monkey females typically do not really enjoy caring for their young. They will haul the infants around, the babies clinging to the mothers' fur; but the mother grows increasingly impatient about the arrangement. Left to handle infant alone without her partner's help, she eventually gets downright hostile, biting at the baby's feet, trying to rub the youngster off against the branches in the enclosure at the Center. She continues to do this even if the baby wails in distress.

The males don't do that. They patiently cart the youngsters with them. In titi families, males actually end up doing some 80 percent of the carrying and cuddling of the young, although Mason and Mendoza describe them as merely more tolerant—they let the little ones hang on—than solicitous. When titis grow up, they leave their parents without looking back, running off to build their own happy coupledom.

"They're not parental," Mendoza says. "They really don't care about the kids. I mean, they really don't. They bond with each other. You don't see them rushing over to the kids when they vocalize, as you do with other monkeys. They spend an enormous amount of time, though, negotiating with each other, monitoring each other. That's the switch. It's the relationship that's so fascinating."

Do they shed any light on humans as monogamous couples? Mendoza adds this perspective: "When talking about physical dimorphisms and how much alike the sexes are in monogamous species, we may employ the wrong lens. Titi monkeys have no problem recognizing male and female and do so instantly. We often don't know except by cultural conventions regarding hair style, clothing, etc. I remember in the 1960s, the common complaint of my parents' generation was that you could no longer tell boys from girls. How different are we really? Probably a bit more [polygamous] than titis, but not as much as any of the truly polygamous species that I'm aware of."

Of course human beings aren't titi monkeys any more than we are peacocks. Up to this point, I've been arguing back and forth between our species' resemblances to polygamous species and its parallels with those that are monogamous. That's not unlike what the research community tends to do. Researchers have a hard time with the question of whether biology has crafted us—whether nature has intended us to be faithful wives and husbands or gadabout sleeparounds.

It would be comforting to find support for monogamy. Monogamy seems, somehow, so much more intelligent and thoughtful, so much more civilized than the screw-you-today, your-sister-tomorrow style of, say, rooster and chicken. But as anthropologist Bobbi Low, of the University of Michigan, says, our marriage patterns seem to suggest that if we are monogamous, ours is at best a form of serial monogamy—an approach in which the first partner might be succeeded by perhaps a better partner and then, perhaps, by a better one. Or maybe we are even less monogamous than that. Low has analyzed human relationships in terms of monogamy in cultures around the world. And her work shows that we fail to meet the high standards of the titi monkeys.

Based on marriage habits, Low classifies humans as "slightly polygamous." When anthropologists catalogue human societies—each separate tribe and culture—they find that the vast majority, some 83 percent, allow for some kind of polygamy. We define polygamy mostly as a system in which males mate with multiple females, the extreme examples for us being harems or the way that high-ranking Chinese men used to maintain a series of wives. The opposite of that—in which females maintain multiple mates—is called polyandry, but Low points out that this is extraordinarily rare and appears dictated by unusual events. In one region of remote northeastern India, for instance, agricultural subsistence is so difficult that it really takes two men to assure that a family's children will survive. In some villages, two brothers marry the same woman, essentially establishing a two-husband household, although the children are all formally raised as the children of the older brother. There have also been cases, Low says, of extremely well-off lineages who wanted to keep and concentrate the wealth and so, again, two brothers both married the same woman.

In the United States, though, she suggests that we maintain a kind of informal polygamy. The overall pattern in America is that more divorced men than divorced women remarry, and when they do remarry, more of the divorced men begin new families. This is overall, of course, and just about anyone can think of an individual exception. But since the general definition of polygamy is a system that provides men, more than women, with multiple partners, it turns out that we in the United States look more like polygamists, in general, than serial monogamists.

As I said, there's a lot of debate over this anthropological viewpoint. If we go back to the purely biological, we can find yet another angle on human monogamy. And, in its own way, this biological approach is just as provocative as Low's analysis of American culture. It says that within animal families—field mice and voles are our examples—you can find that one species is monogamous by nature while another, although extremely closely related, is not. The California field mouse is among the most monogamous animals on the planet, but field mice of the Midwest are nothing of the sort. The prairie vole goes wholeheartedly into partnership, but the montane vole, which likes a hillier terrain, is aggressively promiscuous. And that suggests that monogamy is, in part, a very specific adaptation—it evolves as an exception to the rule because, for reasons not understood, the father becomes essential to the survival of the young.

David Gubernick, who studies the California mouse, points out that this need for the dad's participation could be greatly influenced by environment, or an adaptation to a particularly perilous niche. Perhaps food is scarce and it takes two parents to forage for the young; or a cold climate demands extra body heat from the adults. Or it could be something else. Environment isn't everything. Whatever pushes a species into monogamy, the California mouse demonstrates that it begins a chain reaction that gradually matches the biology to the required behavior.

There seems to be a complex and life-reinforcing chemistry to parenting and partnership. Research suggests that once that route is chosen, and the single life abandoned, the brain itself changes. And this makes sense, of course, because if parents are going to be any good at all at what they do, they have to change in some fundamental ways. To a great extent, you have to kiss self-absorption goodbye, accommodate a partner, and care for a small child whose survival may depend on that commitment.

In a laboratory experiment rife with soap opera potential, Gubernick, who also studies the mice in a spectacular gold and green nature preserve in California's Carmel Valley, pried couples apart. He then offered the newly separate males a chance to cheat without consequence, shielding them from their mates and housing them with fully receptive single females. The males—to the genuine surprise of the scientists—rebuffed their new roommates, turned their backs, and even snapped at the females.

The separated female partners, especially when they were ovulating, were less loyal. When offered an equal chance at free sex, about 15 to 20 percent did mate with an available male. But that still left an overwhelming majority that were faithful. And when Gubernick studied the females more closely, he found that those who did mate with strangers tended to be females who either had trouble achieving large litters or who had recently lost infants.

Unlike that of the titi monkeys, the focus of the California mouse partnership is the children. The fathers are engaged with the infant mice from birth, licking them just as the mothers do, cuddling and keeping them warm. The babies wouldn't make it without Dad; they depend on him for survival. In a classic lab experiment, Gubernick compared ten couples with young to ten single moms, having removed the males from the parenting duo. The couples raised their young successfully; the single mothers failed terribly. Only three managed to keep the baby mice alive. In most cases, about two weeks after the father was removed their milk production simply shut down. "We don't know why that was," Gubernick says. Maybe the exhausted females simply reached their limit and quit, sacrificing their young for their own survival. The baby mice are too tiny to keep any kind of body temperature going, so in a normal family, the dads cuddle the babies and keep them warm while the mother is out foraging for food, and vice versa. In fact, the father is so vital to survival that you can argue that this explains their sexual fidelity—that natural selection would have favored males with a bent toward faithfulness.

There's some remarkable chemistry that underlies all of this behavior in the California mouse. For instance, Gubernick has demonstrated that the female uses a kind of chemical signaling—an as yet unidentified compound in her urine—that induces male parental behavior. He and his colleagues found that they could induce parental behavior in most of the male mice by dabbing female urine on their noses (which certainly takes the romance out of parenthood). Without that whiff, or preferably the female herself, the males were distinctly less cuddly and more prone, actually, to attack the infants.

But—and this makes an interesting parallel to humans—there was remarkable individual variation. In fact, these experiments accidentally uncovered a group of males apparently born to parent. In the first study of urine exposure, Gubernick found that between 26 and 45 percent of the male mice showed parental behavior even when they were dosed only with distilled water. They were just born nice, cuddly guys. Partnership only improved on that. However, there were other males who, no matter what the relationship, continued to attack the infant young.

In a later study, Gubernick compared a group of bachelor males who were infanticidal to those who were less baby-hostile. Mice from both groups chose a mate and helped raise their families. But their styles were consistently different. If the female (and her nurturing influence) was removed, the gentler male mice stayed that way, even without her chemical signaling, and they continued to care for the young. The other mice, lacking a female partner for several days, reverted to infanticide. "So a small majority became permanently parental, even without the female," Gubernick says. "What's different? We don't know. We're looking at their brains now. But we have no answers, yet."

Remember the preoptic area of the hypothalamus, where so many scientists have looked for gender differences in the brain? That's exactly where Gubernick began his search for understanding parental behavior in the California mouse. And there, he made a fascinating discovery. Before mice become parents, the preoptic area is distinctly different in males and females. The region is consistently bigger in the males, averaging more than 6,000 or so neurons, compared to about 4,500 in females. The individual male neurons tend to be a little chunkier, too.

That abruptly changes when the babies are born: the female neurons start getting bigger and the males start a cell countdown of sorts, their preoptic areas getting smaller. By the time the males are routinely warming the infants, the preoptic areas of males and females look virtually identical. In part, this shows once again how difficult it is to nail down a brain structure difference: study the mice as virgins and they'd be sexually dimorphic in the preoptic; study them as parents and they wouldn't. But the implication is that parenthood is a great equalizer for the mice; they share in the responsibilities and their brains become more similar.

"Is it that they're acting parentally and so the brain changes?" says Gubernick. Or is there some hormonal change, related to parenting, that drives the process. Or does one precede the other—act like a parent and you produce parental-type hormones? "We don't know," Gubernick says, "but the exciting thing is that these changes occur, that we've found them, and that we know they're real."

There's also some very distinct hormonal chemistry associated with being a good partner. Many scientists place two particular hormones—oxytocin and vasopressin—right at the heart of monogamous behavior, although other hormones, including testosterone and the estrogens, may play a supporting role. Researchers suspect that oxytocin and vasopressin play an influential role in romantic love and parental love. The fact that there are two of them also suggests a significant gender difference. Both sexes make both hormones. But oxytocin is the hormone linked to emotional connection in females. In males, it's vasopressin. And one of the mysteries of monogamy is why the sexes might need two different hormones to achieve the same things—a sense of partnership, the urge to care for a child.

That difference is dramatically illustrated by comparing two species of wild rodents: the prairie voles, dedicated and parental homebodies, with the polygamous and restless montane voles. The bottom line is that the oxytocin and vasopressin run high in the monogamous voles, rising as they pick a partner, and rising again with parenthood. There's no comparable chemistry in the montane voles, who move from mate to mate. Montane vole females raise their infants alone.

Thomas Insel, a neurobiologist who directs the Yerkes primate center in Atlanta, and his colleagues can trick female prairie voles into bonding with males by a single oxytocin injection into the brain. And male voles given a similar shot of vasopressin hurried to cuddle next to a female, even if she was ovariectomized and there was no sexual intimacy. Another study, at the University of Massachusetts, found that blocking vasopressin in male prairie voles caused a dramatic drop in their parental behavior; they were three times less likely to stand protectively by the young or to help warm them.

But the hormones are clearly potent enough to influence social behavior in polygamous species as well. If scientists block vasopressin in male rats, they lose much of their social network sense. Time and again they have to resniff and relearn their relations with other rats. And scientists can make a virgin female rat act like a mother with a single shot of oxytocin into the brain. This is a bigger deal than it may sound. Most virgin female rats are vicious with the young of others. But add in oxytocin, and they'll cheerfully build a nest for adoptive young, and warm them and clean them.

In mammals, including humans, oxytocin and vasopressin are made in the sexual center of the hypothalamus. We don't understand them well in humans, partly because, like everything controlled by the hypothalamus, they're required to be multipurpose. In women, for instance, oxytocin triggers uterine contractions in childbirth (a synthetic version, pitocin, is often given to boost labor). It also stimulates the "let down" of breast milk to a suckling child. In men, vasopressin helps keep the water-mineral balance in the body (it is often called an antidiuretic hormone), stimulates blood vessel constriction, and helps control blood pressure.

Beyond that, the evidence for direct influence on behavior is pretty thin. In men, vasopressin rises when arousal occurs. If arousal progresses far enough—eventually to orgasm—oxytocin floods out in both men and women. There's no direct evidence so far, though, linking these hormones and human behavior. One Swedish study found that nursing women, directly after childbirth, were measurably more responsive and sensitive to the feelings of all around them than they were before birth. This could be linked circumstantially to a rise in oxytocin. Researchers see that as a genuinely provocative finding but not as proof.

Gubernick did test for the possible influence of oxytocin on his male monogamous mice, but found no significant connection. Oxytocin was higher in expectant fathers than virgin males. However, it started declining before the litter was born and continued to drop during the period that the females nursed the young. Further, among expectant fathers, there was no difference in oxytocin levels between the baby-killers and the nice guys.

Instead, Gubernick discovered a different hormone rising in male mice: prolactin, which is made in the pituitary gland. In the females, prolactin is the hormone that stimulates milk production. Rising estrogens stimulate the prolactin-making region of the pituitary (called the lactotroph) so much that the whole gland gets visibly larger during pregnancy. In men, the pituitary also makes prolactin, but its role is less clear, although there's some evidence that it can indirectly influence testosterone levels.

As the California mice moved into fatherhood, the prolactin surge was so strong—so compelling even—that Gubernick began to wonder whether it might make a difference in humans. He set up a small experiment in which he tracked prolactin and testosterone in nine men, all expectant fathers. He tested for the two hormones for eight weeks—four prior to childbirth and four afterward. There was, disappointingly, no real fluctuation in prolactin. But the testosterone shift was unmistakable. It fell in all the men, sliding steadily downward almost immediately after the child was born. Says Gubernick, in puzzlement, frustration, and fascination: There needs to be a study that looks at a suite of the hormones in men —prolactin, testosterone, oxytocin, and vasopressin—tracking them from expectant parenthood through delivery.

The testosterone finding is still interesting on a stand-alone basis. For example, there's some evidence that levels of testosterone can predict whether a father is going to be there for his children. John Wingfield at the University of Washington in Seattle has demonstrated this with real precision. His studies of those stars of monogamy—the birds—could convince anyone that Darwin wasn't so far off, after all, in making his bird-to-human analogies.

As previously mentioned, birds are a showcase of monogamy in all its variations. There are different levels of commitment, of course—from shorebirds, such as petrels, who tend like the mice to be faithful partners, to barn swallows, who perpetually seem to be just checking out the possibilities. In general, high testosterone birds are absentee fathers and dedicated polygamists; in some of these bird societies, the battle to mate is all-consuming—and all-important. Among species with what Wingfield describes as "screamingly high testosterone levels," a single, super-macho, alpha-type male may account for 90 percent of the young in a single breeding season. But he's so stunningly successful only for a very brief time, a true live-fast-and-die profile. Most of the "screamers" last about two breeding seasons tops, before the hatred of other males runs them off, or they drop dead of infection (since high testosterone can shred the immune system) or, possibly, of exhaustion.

In the bird world, Wingfield suggests, there's a behavioral loop, something about parenting that tends to keep testosterone down so the father can focus his energy on the offspring. And this is reinforced by the habits of monogamy. The more peaceful relationship reduces the constant prodding of the testosterone system (biologists call it the "T system") to brace for aggression. As the hormonal production settles down, the male settles into domesticity. Monogamous species such as the sparrow are essentially monomorphic not only in outward appearance but also in their circulating testosterone levels, which are basically the same between sexes. Female birds in monogamous species tend to have higher testosterone than females in polygamous species.

There's a final, cautionary point about what monogamous birds and their approaches to parenting might tell us: Here you have an evolved system of nurturing partnership, an evolutionary arrangement of task-sharing so influential that even some primary biology, such as testosterone level, becomes nearly identical between the male and female. Does that guarantee total sexual fidelity? No. And does it create entirely gender-neutral performance, in which Dad and Mom interact with the chicks in exactly the same way, given that they are both so focused upon raising their fluffy young? The answer is no again. Monogamy—and this may ease some misgivings as we freely draw parallels between bird species and the evolutionary possibilities of our own—has yet to turn males into females. We're talking partnerships here, but we are not talking indistinguishable androgyny. In one of my personal favorite experiments, evolutionary biologist Judy Stamps proved this by studying a very familiar bird, the household parakeet, known popularly in Great Britain as the budgie.

Budgie males and females build a cozy partnership. When the eggs are laid, the dad bustles about, seed hunting, bringing back choice meals to the female, and watching over her while she warms the nest. When the eggs hatch, the two parents share in feeding the young. But Stamps found some very clear differences in how mother and father approached the feeding of their infants when she observed their behavior after the babies hatched.

Budgies are unusual in that there's no simultaneous hatching of baby birds. The eggs are laid in a sequence and they hatch accordingly, so that the nest ends up with a family not unlike a human one—a bunch of children of varying ages, from older, tougher nestlings to tiny, helpless newborns. When the researchers first began observing the budgies, they assumed, basically, that the infantile lastborn was doomed—that it would be crowded out by the pushy older siblings.

Not so, thanks to the scrupulously fair budgie mother. The researchers found that parakeet mothers feed the chicks according to need. It didn't matter if the older nestlings screeched and pushed; the mothers in the study started by making sure the youngest got his or her share, and then worked upward to the biggest and loudest.

Now, the male birds really wanted to help feed, too. In fact, even though the largely nestbound mothers continued to beg for their own meals, the males could be seen edging around the mate, eagerly delivering carefully chosen seeds to the infants. But a father approached the task with less fairness. He tended to be overwhelmed by begging, so that he stuffed the seeds hastily into the loudest mouth. After one meal with Dad, the youngest birds came to learn that begging worked, so the whole nest became incredibly noisy as all the little nestlings screeched their fuzzy heads off in the hopes of influencing food division. "It became deafening," Stamps says, laughing. Her point, though, is more serious. Even in this committed partnership, the female budgies spend more time with youngsters, know them better, and can gauge the needs better. And the males are less sure of the proper thing to do, possibly because they spend more time hunting up food.

A scientist like Stamps has to know the species to understand what makes a good parent, male or female. She's learned that she can't expect the definitions of the two roles, even in the most parallel of partnerships, to be exactly the same. We can't always predict that the mother is the best caretaker, as exemplified by the titi monkeys. Behaviors are incredibly specific to species, time, and place, she points out. We tend to view this all through our own lens: it would be nice, or at least politically correct, to find evidence that a good dad should act like a good mom, and that females are the gold standard for parenting behaviors. But it ain't necessarily so.

In some species, a father's most valuable contribution may be to get out of the way, Stamps says. The dad may help simply by not competing with his offspring for the available food. Or it may be that guarding the territory, fending off predators or even other, infanticidal males, is a lot more important than staying home and cuddling the young.

And what about us? We know children, lots of them, grow up without fathers. And we try, as a society, to provide safety nets for the children of single parents. But at the basic, precivilization level, was there an essential need for the father to be there if the child was to survive? Do we have a biology of our own—like birds and mice and monkeys—that's part of partnership? We share the same hormones, of course, but whether they work the same way in us, we don't know. Thomas Insel, of Yerkes, has been looking at monkeys—a relatively closer parallel to humans—for an oxytocin/vasopressin receptor system in the brain similar to that found in voles. So far, he hasn't been able to find it. The patterns look different, he says, and his laboratory is still trying to figure out what that means. The first answer, he emphasizes, is that we don't yet understand the chemistry of partnership and attachment in primate species like our own.

We're a fascinating puzzle in the monogamy story: we bond and break the bonds; we promise fidelity and cheat on the side; we lack the perfect body match of wholly monogamous species and yet many of us believe in a holy state of monogamy. The fact that our bodies are not monomorphic supports that point. I've come to think of humans as "ambiguously monogamous." If we lack the uncompromising dedication of the decisively monogamous species, it's interesting to propose that it's because we're in a state of transition, moving slowly away from the more polygamous habits of our ancestors.

"Our culture tells us that equal role division is the thing to do, but biology tells us that may not be the history of our species," says Judy Stamps. "It's not politically correct to say these days, but there may be differences in aptitude for child-rearing that come with that." Stamps points out, however, that the fossil record suggests a slow evolution toward greater monogamy; some 300,000 years ago, early hominid male skeletons were nearly twice the size of female skeletons. Among modern humans, the difference is closer to 17 percent. "My suspicion is we're in the process of going in that direction," she says.

Stamps has a typical narrow slice of an office at the university where she works, filled with sunlight, books, papers, coffee cups. We're sitting in the midst of this cheerful clutter (I always think of clutter as cheerful, because otherwise I would have to think of my house as a mess) when she raises this point. And it's *my* house, actually, that comes to mind as she talks—or at least, the way my husband and I try, in our marriage, to find a fair balance in caring for home and children. I start wondering out loud about this, putting it to Stamps as an argument of sorts.

At my home, we have a fairly common modern agreement in which, if I cook, he cleans, and so on. It's the same with many of our friends—all of us raised during the baby boom with the *Leave It to Beaver* paradigm, so it's not always easy. We try, pushing a little and giving a little, to achieve some measure of fairness. But could we push a lot harder? And if we did, could we achieve an equal partnership that would force the human race to change?

Could we somehow kick our species' evolution into overdrive on our journey toward true monogamy? It's really no more than a thought experiment, poised on the too-easy edge of fantasy. Suppose we, as a human species, decided to put equal partnership before all. Suppose we shared in everything. Suppose we adjusted constantly to our partners' needs, with the fairness meter ticking at all times—an "equality clock." Over time, would we become more monomorphic? Would we be big-brained titi monkeys without fluff-covered skin and plumy tails?

Not unless I could talk a huge portion of humanity into joining me in the quest, and that's not likely. As Stamps reminds me, my marriage and those of my friends follow a pattern for educated, middle-class American couples. It's nothing like a worldwide majority trend; there's no great unified rush toward complete equality. That doesn't, and shouldn't, rule out more equality as a goal. But I know that in my lifetime, and over many lifetimes to come, the human species will remain kind of ambiguously monogamous. There's just no real evolutionary pressure, Stamps says, for a specieswide change.

On the other hand, there's an equally interesting smaller question. If I focus only on my own home, where the fairness meter ticks pretty steadily, then there's also the issue of a more personal biology. If I, as half of a two-career, two-child partnership, emphasize equality as does my husband, does that change us? If I'd spent my marriage acting like June Cleaver, would I today be wearing pearls as I dusted doorframes and—apparently—liking it? Or, to be less flippant, if I'd spent less time in the power field of journalism, would I relax more easily into domesticity? If my husband, in turn, didn't give so much of himself to the children, as he does, would he maintain an edgier and more competitive biology that would push him to spend longer, more aggressive hours at the office?

Marian Diamond is the Berkeley neuroscientist who literally expanded our realization that the brain never quits responding to new events and opportunities. So when I started pondering this idea, I called her up, and, to my pleasure—because I like the idea—she also saw the possibility: "People do wonder about what's happening in our society. Are we going to lose masculinity by men taking care of babies? And will females gain aggressiveness, beyond attractiveness, by becoming corporate leaders? I think that can happen. We're products of our environment and, as it changes, I can't help but think the brain will change too. It depends on the balance. If a man spends one to three hours a day at home, is that really going to change anything? Maybe, maybe not. But if he's involved in the home-keeping full-time, if he's really there? Probably, yes. The main point is, the brain is so flexible, it can change at any time in a healthy individual."

If we are really traveling toward monogamy, we might be expected—like titi monkeys—to pick mates on the basis of their "good partner" characteristics rather than their "good looks." And there is some evidence along those lines. When David Buss, one of the best-known researchers of sexual evolution, conducted a 33-country survey of what men and women wanted most in a mate, both sexes rated "kindness" and "intelligence" at the top of the list. Buss concluded that the high kindness rating particularly emphasizes the priority we put on partnership. In another study, he found that while women saw dominant, high-status men as sexually attractive, they didn't find them so appealing as companions—giving them low marks on warmth, tenderness, and likability.

That's not to say we've all become touchy-feely in choosing our roommates. Or that men and women—by some unlikely stretch of the imagination—look for the same qualities in their partners. They absolutely don't. Kindness and intelligence were first on both lists. But after that, the lists diverged. Men rated youth and beauty as next in importance; women rated wealth and indicated a preference for older men.

Does anyone doubt that women's looks are of major importance to men? When researchers from St. Edward's University in Austin, Texas, placed "test" personal ads by two imaginary women, one who was a recovering drug addict and the other 50 pounds overweight, 80 percent of the total respondents replied to the addict. On the other hand, a joint study of personal ads by scientists at the University of New Mexico and the University of Liverpool found that women were five times more likely than men to request in their ads that a prospective partner have money. We can, and we do, get very cynical about these patterns: women are gold diggers and men want women as trophies. But, whether we like it or not, there may be something of a rational biology in those choices, if indeed our ancestry is a combination of commitment and promiscuity.

Remember, for instance, that our male ancestors may have used physical appearance as a guide to judge a healthy mate. That approach may have been handed down preferentially in our evolution. And remember too that in monogamous species, females are drawn to the male with the most resources; in titi monkeys, for instance, the lure is a good territory. A wealth of resources—say, a strong food supply—can define a mate who will give offspring a better chance of survival. It provides key information to a female looking for a good partner, especially if keeping the offspring alive requires two contributing parents.

It's true that our children can survive with only one supportive parent. But can they thrive in the same way they do with two? This is not a slap at single parenthood, which happens for many reasons. Often those reasons are both admirable and brave: it takes courage to leave a vicious or abusive spouse; it takes strength to build a life after losing a good spouse to death. It's also true that many single parents do a terrific job of raising terrific children. But although I've discussed how necessary a giving mother can be to proper child development, there's a new body of evidence showing that the give-and-take between father and child encourages different but also important strengths. Overall, fathers rough-and-tumble more with their children; they tease them more; they play harder and tend to emphasize the physical challenge more. They tend to comfort less than the mother; they are less prone to try to cushion a child against the world.

Development specialists, such as Jay Belsky at Penn State, or Ross Parke at the University of California, Riverside, say research suggests that such challenge, even such less-sympathetic handling, has real benefits. Fathers push the child, and even if the push is too hard, that's not catastrophic. Children need to learn to handle emotional ups and downs, Belsky says, and fathers, by keeping the children more off-balance than mothers, help them do that. Curiously, because fathers may be less "there" emotionally—less open—than mothers, time with fathers also teaches children about reading subtle interpersonal cues: Is he teasing or serious? Is it in his face? The tricky part, Belsky says, is balancing that against a child's need for affection, and not pushing a child so far he begins to feel alienated. Because of this, mothers, with their more cuddly approach, often manage to forge a better long-term relationship with the child.

There's one more aspect of fathers as involved parents that is worth considering. The point comes from a study out of Israel that looked at whether fathers spent any time at all with their children. The study began: "It is well known that fathers, on the whole, are much less involved in infant care than mothers, be it in terms of responsibility, time spent in interaction or performance of caretaking abilities." Remember the issue of whether partnership had advanced further in Western cultures? Well, the researchers found that the Israeli fathers who spent the most time with their children had gone to school or lived in an industrialized Western country. But even adding in those dads to the study pool, the overall track record was still dismal. The researchers interviewed 160 fathers. On average, even with full-time jobs, they had available about three hours a day that they could spend helping to care for the baby. Their actual participation, though, averaged only about 45 minutes a day. The survey measured ten caretaking tasks (changing diapers, feeding, and so on); on average, the fathers performed only one such task every day, with more than half of the men interviewed doing none.

Obviously, this is not equal-parenting partnership. Neither does it come close to describing how much time many fathers spend with their children. Just as Gubernick found mice to be either naturally loving fathers or reluctant ones, perhaps the same is true with humans. Or perhaps this study overemphasized the more indifferent men, tending to focus on those who believed that a mother's place was with the child.

But, setting aside that male-female power balance for now, let's talk about what such indifference means to the child. The study's authors found that fathers who spent more time with their babies —who were nine months old at the time—actually liked the children better. They saw them as smarter and, as a result, probably helped them get smarter. Many of them sat with the child in the evening, reading board books. Fathers who spent almost no time with their infants tended to think of them as still too dim to grasp the idea of a book. We can follow this with another study, also out of Israel, looking at three-year-olds. It found, again, that fathers involved with their children tended to perceive them as more capable and brighter. In turn, the children of involved fathers tended to perform better at preschool than those of bored ones. And, in turn again, the fathers who spent more time with their children expressed more pride and more pleasure in them.

If we want to define ourselves as a monogamous species, then we need to accept all the biological implications. In many such species, the father's participation is essential to raising a whole and healthy child. That doesn't rule out exceptions. It just means that some standout monogamous species—mice, voles, monkeys, us (if we agree)—have a biology best suited to being raised by two parents.

I'm not proposing some biological endorsement of the modern political emphasis on "family values," which seems often to lean heavily on the men-as-breadwinner and woman-as-homekeeper stereotypes. If we believe in human monogamy, I think, then we should also believe that it requires a fair partnership, so that both of us, men and women, bring resources, love and affection, and support to all aspects of marriage. I don't think we're there yet—

not planetwide, anyway. But I do hope we're moving in that direction. The benefits of equal partnership seem enormous both to couples and their children. Perhaps we need not worry about the big evolutionary picture, only about passing that lesson on to the next generation.

Chapter Five

THE SECOND DATE

Inquiries into Sexual Orientation

The most consistent behavioral difference between men and women is their sexual orientation. Almost always, men are attracted to women and women to men. That sense of attraction is so intrinsic that we don't even wonder at the response. It's like sweat, like the way our eyes shut when we sneeze, and like the ebb-and-flow of air in our lungs. Sexual orientation is basic.

In the least scientific of ways, you can test this by asking anyone to remember the moment at which he or she consciously chose an orientation. It's ridiculous. What teenage girl decides, formally, that she's going to giggle over boys? What man sits down at the kitchen table with a tablet and pencil, makes a list of pros and cons, and says, "By God, I think I'll orient myself toward women"? It makes more sense to wonder, instead, when that involuntary lust for one sex or the other begins, such that our breath catches on the easy flex of muscles or the swing of a walk.

I've been describing, obviously, orientation to the opposite sex. The same principle of instinctive sexual pull would apply to same-sex orientation as well. My point is not, at all, to put homosexuality on some separate plane. Rather it's to reemphasize how universal sexual orientation is—how and why it develops is a question for all of us.

These days it seems the term "sexual orientation" is used almost exclusively to refer to same-sex attraction. If I pick up a book on the biology of sexual orientation, it is almost certainly going to be a lengthy discussion of why people "become" gay. I'm not sure how we arrived at this point in public consciousness, because when you think about it, what makes same-sex orientation fascinating from a scientific point of view is that it's rare. The best numbers for the United States say that practicing homosexuals make up about five percent of the population. Michael Bailey of Northwestern University, one of the most respected researchers on the subject, thinks even that overstates the case.

Bailey suggests that the real percentages of those practicing same-sex behavior—people involved intensively and exclusively with sexual partners of their own gender—are closer to two to three percent of the male population, and 1.5 percent of women. Statistics are slippery things, so let's clarify these. We're talking about men who sleep with men and women who sleep with women on a regular basis. That's different from experimenting once or twice. But if you add the occasional one-night adventure, it pretty much doubles the numbers, up to six percent for men, three percent for women. Then, if you count everyone who ever let curiosity flicker across the mind, ever wondered what it would be like (and admitted it to a survey taker), then the numbers climb again, to about 8.7 percent of men, 11.1 percent of women.

For political purposes, those differing groups tend to get mixed together into the notion that close to 20 percent of the population is either homosexual or would like to be. I understand why those working for gay rights would want the numbers to appear as large as possible. We live in a culture that is needlessly and pointlessly hostile to homosexuality, so advocates want a larger, and thus a stronger and more influential, constituency. But I think overstating the case is a mistake. It makes it harder to sort out what's real.

If you ponder the real numbers fairly, they offer some valuable insights. Many more people consider sleeping with someone of the same sex than actually do it. Women, especially, seem to call up these fantasies, dream a little about other women, and then fail to follow through. Almost all people who study sexual orientation say that women have a more fluid sexuality and are far more capable than men of being aroused by either sex. I'll pursue that point in more depth later.

The fact that desire can spark without catching fire may mean that we're tracking no more than passing curiosity or a random flutter of hormones in the blood. It may say something about the power of social rules to suppress lust. Perhaps, as advocates suggest, some of these sexual dreamers would follow through if they weren't so afraid of being castigated for it. But no matter how far you stretch the percentages, stacking together thinkers, adventurers, voyeurs, and confirmed gays, you still end up with a minority—albeit a vocal minority with legitimate concerns.

The really fascinating stuff about sexual orientation starts with the fact that we all have one. Whether the compass orients to the traditional north or quivers away to the south, there is a basic pull involved. We are all moved by human sexual magnetism of one polarity or another.

"You get the sense from some people that this kind of research [into sexual orientation] is not serious. That's dead wrong," Bailey says, sounding both harried and worried. "I hope people who make the decisions on what research to support don't think that way. I get the sense that people think that, at the worst, studies of homosexuality are Oprah-ish, and at best only relevant to an unusual population. [People think] that we can't find out about human nature from such studies. That's simply not true. Almost everyone has a sexual orientation toward either a man or a woman. Finding out how that develops will help us understand a basic part of who we are and the difference between men and women."

This research is not some kinky little exploration into a strange and limited netherworld. The fact is that researchers really don't fully understand how we, or how other animals, develop sexual orientation. We know that in other species, same-sex orientation is also rare, but far from unknown. Some form of homosexual behavior—be it only that passing flicker of interest—has been observed in more than 60 species. There's even some evidence of long-term same-sex orientation. Female Japanese macaques, a species of Asian monkey, will build a consort relationship with other females for years running. And some rams in a number of sheep species exhibit a dedicated passion for other rams. You can hardly argue that some perversion of sheep culture turned them that way. We don't know what creates a ram-loving ram and we don't know what causes a man to be gay.

Does sexual orientation begin in genes? Is it sculpted by steroid hormones, like so much of the reproductive system? Can environment alter its path? Do we get to choose our orientation? (That's a different question from choosing the person who shares the bed, since we can decide to sleep with someone who doesn't overwhelm us with lust.) And when does this start? At what stage of life are we officially oriented? Should we expect it to come with the sexual wash of hormones in puberty, or long before?

It's almost startling to realize how much of this field is made up of questions that don't yet have satisfactory answers. Is sexual orientation simply a facet of desire, tucked into some reproductive compartment of the brain? Or does the whole brain orient, to some degree, with that compass needle? Do you accept—as some evidence indicates—that a homosexual man possesses a brain organized, overall, more like a woman's? If that's true, is this organization caused by hormones or has the brain merely responded, or built structures, to accommodate the man's experiences and his lifestyle?

We know same-sex orientation has been with us a long time. Unmistakable references to homosexual partners are laced through recorded human history—in the Bible, Egyptian hieroglyphics, Greek and Roman poetry. Study of cultures primitive and modern, ancient and contemporary, tells us that same-sex lovers have never been a passing fad; they are not unique to either a particular way of life or a particular period. But from the standpoint of simple reproduction, we have to wonder why this is. If same-sex orientation is, indeed, woven into our evolution, what explains its conservation? Wouldn't one assume that, from the point of continuing the species, this would be a lethal behavior? If it's in the genes, how do those genes get passed on?

The last question leads directly to the risks of oversimplifying and overestimating genetics. At one level, logic might suggest that any such gene would be knocked out as fast as possible. Actually, evolutionary theory can stretch to accommodate homosexuality. Sociobiologists such as E. O. Wilson have suggested that while such behaviors might not advance the individual's genes, they could help promote kin survival. The idea behind this is that perhaps homosexual men helped care for children, protected family members, and thus aided survival of their kin. Anthropologists have traced those very behaviors in a number of cultures. They've documented, for instance, among early American Indian tribes gay men who would work with the women while in the villages, and also do housekeeping chores on hunting trips, while other men tracked and killed game.

I have difficulty with this caretaking concept, though. I can't quite conjure up enough faith in the idea of evolution making a special place for homosexuals, just as long as those homosexuals were altruists. What if they weren't?

Harvard anthropologist Stephen Jay Gould did a pithy analysis of the "helper-giver" theory, describing it as well-intentioned in trying to accommodate homosexuals, but also patronizing. In the end, he suggested we might try less hard to account for every scrap of genetic influence and try harder to give people some privacy in the bedroom.

Of course, homosexual-as-giver isn't the only genetic theory going. Some researchers suggest that perhaps the homosexual traits are bundled with other genes that confer reproductive advantages—that the gay genes ride piggyback, if you will—and that we simply carry them along from generation to generation. The genes could be continued through families or individuals. Many gay men and women do have children, encouraged or even intimidated into a heterosexual lifestyle, including reproduction. Finally and obviously, homosexuality has been around for countless generations without its nonreproductive aspect making a dent in the unbelievable flood of humanity across the planet. In fact, considering the march of human population—some six billion and counting—I could make the argument that the planet would be a little healthier if we had more same-sex couples and fewer heterosexual couples busy pursuing their reproductive potential.

"The essential genetics may not directly code for homosexuality at all, but something correlated with it," Bailey emphasizes. "Something that's advantageous. What is it? We don't know. The alternative idea is that it's simply darned hard for biology to guarantee heterosexuality every time, that it's not a stable system. The problem with that [theory] is that if it's hormones that set sexual orientation, they don't seem to have much problem guaranteeing that men get penises. So, why can't they keep sexual orientation straight? On the other hand, homosexuality is very rare . . . in other words, we don't know."

It was Bailey, with colleague Richard Pillard of Boston University, who set off today's zealous hunt for the genetics of sexual orientation. They put together a series of studies that almost everyone agrees established that there's a genetic "something" in sexual orientation. "Everyone likes to nitpick," says Daryl Bem, a psychologist at Cornell University. "In the end it comes down to whether you believe the data or not. I believe the data. And part of that is that I trust Mike Bailey. He's very honest about what he has and he's very cautious in interpreting it."

In the early 1990s, Bailey and Pillard published a series of studies of twins, based on interviews with gay and straight brothers. There's a solid logic to twin studies: basically, people produce two types of twins—monozygotic (one egg, split) and dizygotic (two eggs, hanging out together). Most of us call monozygotic twins identical and dizygotic twins fraternal. The difference is more complex, and more interesting, than whether the twins have matching faces. Because they come from the same egg, identical twins get identical genetic material—barring, say, the occasional mutation. Fraternal twins, from different eggs, are as genetically close as any other siblings—

about a 50 percent match. But, like identical twins, they share what scientists call a "twinned" environment. They develop in exactly the same amniotic fluid, equally exposed to whatever the mother eats or drinks. They age at the same rate, playing more closely than siblings separated by many years. Identical or fraternal, they are treated by others as a unit in the way that other siblings are not. If you want to search for heritable influences by comparing the tightly matched genetics of an identical twin to the standard genetic link between siblings, fraternal twins are the best way to do so. They let you filter out environmental interference.

Bailey and Pillard recruited 110 pairs of male twins, half identical, half fraternal. In each case, they knew that one twin was gay. They then sent a questionnaire to the other brother in each pair, to determine his sexual orientation. Among the identical twins, 52 percent of the brothers were gay. Among the fraternals, the number was 22 percent, high enough above the background population rate to suggest that there was something distinctive in those families. The researchers found a very similar pattern with lesbians.

And Bailey has looked for confirmation abroad. His recent study out of the Australian Twin Registry, with almost 5,000 participants (roughly 1,800 sets of twins and 1,300 unmatched twins), also tracked the same pattern. Bailey is quick to emphasize, too, that his initial study wasn't the first along these lines. A somewhat informal study in the 1940s, in which the researcher persisted in calling his subjects members of the "underworld," also found a very high probability that if one identical twin was gay, the other would be as well.

Still, Bailey worries that the survey methods—he and Pillard advertised for participants through gay newspapers—may have produced slightly inflated results. That is, people who read advocacy newspapers, who choose to respond to a publicly advertised survey, who enjoy the scrutiny, who like to call attention to their lifestyle whatever it may be, may not reliably represent the entire community. That was one reason why he turned to the broader-based Australian study—and was reassured by the similar results.

Long before that, though, other scientists were taking such surveys very seriously. Geneticist Dean Hamer says he began his own well-publicized hunt for a "gay" gene after reading Bailey and Pillard's work and becoming convinced that there really was something worth seeking. Hamer, a scientist at the National Cancer Institute (NCI), called both Bailey and Pillard and discussed the work with them. He invited Bailey to his home for further discussion before launching his laboratory into a definitive search along the arms of the X chromosome.

Why the X? Other studies had shown that it wasn't only fraternal twins that had a 22 percent correlation for gay brothers; the same connection existed for nontwin siblings—brothers born years apart. In other words, all the studies pointed to genes passed on within a family. When researchers pursued this pattern further, they found that gay men tended to have unusually high numbers of gay cousins and uncles. And they found that these cousins and uncles were almost always on the mother's side of the family. If there was a gene of some sort involved, that strongly suggested that it came on the X chromosome, down through the female line of inheritance, since men get their X chromosomes only from their mothers.

Hamer took the next logical step. He gathered together 40 pairs of gay brothers (siblings but not twins). He took blood samples from each, and sifted out the DNA. Then he and his NCI colleagues picked their way along the long arm of each X chromosome, looking for a place of common ground, a genetically distinct region that would bind these men together. The resulting paper, in the eminent research journal *Science*, reported the discovery of just such a stand-out location, named for the chromosome (X), the arm of the chromosome (q), and its position there (28). A distinctively marked region, Xq28 sits at the very tip of the long arm of the X chromosome.

In his initial publication, Hamer reported that 82 percent of these gay brothers looked alike along that particular region of the chromosome. He had not found a gene, but he had found a startling similarity—one that suggested a gene was there. Could we argue that this was the gene for sexual orientation, imposing some iron biological grip? Hamer doesn't. He says the real significance of his discovery is that it opens up possibility—it shows that we can locate a genetic link to sexual orientation. Finding the location isn't finding the gene, only a hint that an influential one might be there. If it's found, Hamer predicts that it will be only the first of a series of genes related to sexual orientation. Such behavior is too complex, he thinks, to be controlled by a single gene. And if there is an all-important gene at Xq28, he doesn't know its precise purpose: "First we have to find it. It could be anything from a gene that exerts a direct effect on a nucleus in the brain to one that's very indirect, something that affects people's willingness to experiment."

Nevertheless, to the dismay and exasperation of many scientists, the popular news media reported this flatly as the discovery of a gay gene. Some misleading and incomplete reports suggested Hamer had determined that sexual orientation was somehow as genetically fixed as eye color. *Time* magazine's headline read: "Born Gay?"

A number of gay activists, justifiably weary of being accused of deliberate perversion, also played the report that way. The argument that they had simply been nailed by their genes seemed a potent, indisputable confirmation that homosexuality is not something you choose. Hamer has grown weary of the appropriation of his work for political ends. In an ideal world, he argues, it shouldn't matter whether there's a biology to sexual preference or not; we should merely respect each other. "I think it cheapens the science to do it on political grounds," he says. "And I think it cheapens our political system to make decisions based on whether there's a biology to behavior or not."

It's also true that Hamer's work is still new, still unconfirmed. What he has found suggests a possibility, not a certainty. Some other scientists trying to repeat his finding—and in science, independent confirmation is absolutely essential—have been unable to do so. Geneticist George Ebers, from the University of Western Ontario in Canada, did a somewhat similar analysis, although he did not distinguish between bisexuality and homosexuality. He found only a 50 percent correlation with Xq28, nothing more than chance.

Hamer, along with a group of highly respected geneticists from around the country, refined his first work and published again in 1995. This study, which did not make the cover of *Time*, attempted to answer questions raised by the scientific community. Researchers had complained that Hamer's first study failed to compare gay men to either heterosexual men or lesbians. This time, he looked at all three groups.

In the repeat, the genetic link among gay brothers was still impressive—67 percent—but not as dazzling as the original 82 percent. "Both are statistically significant," Hamer emphasizes, but in terms of simple numbers, the second finding appears to leave more room for environmental influences. Straight men did not have the "gay" marker set at Xq28, but they shared another marker. Hamer believes this means that the same gene is active in all men, and that he is looking at two varying forms, depending on whether a man is heterosexual or homosexual.

There was no apparent connection for Xq28 in lesbians. Hamer considers that nonfinding one of the most interesting points. It suggests to him that a different genetic scenario operates in women—that we are looking at two separate sexualities in males and females, governed by separate genes. One of his current projects involves combing the X chromosomes of lesbians and heterosexual women for a different marker pattern.

"It [sexuality] is such an obvious difference between men and women, such a transparent biological difference," Hamer says. "Even the most ardent feminist or masculinist would have to acknowledge that there's some biological input here." Even before the study, he said, he suspected that Xq28 would not appear influential in women: "I think of male and female sexuality as so different," he explained. "You wouldn't expect males and females to share the same gene for testicular development, or for milk production, would you? And this is like that."

And yet, if the gene (or genes) at Xq28 has (or have) been conserved over generations on the X chromosome, then clearly women inherit the gene as well as men. A sister would get the same set of genes from her mother as a brother. What does that gene do in women? Hamer would like to pursue that question, too, in further studies. He finds it fascinating. His speculation is that perhaps in women the gene helps give the perception of men as sexually attractive. "In an evolutionary sense, if a woman was very attracted to men, she'd be pretty hot to trot, right? She might have a few more children and that would mean her genes would be passed along more effectively," he says.

Hamer's work provides a wonderful example of both the possibilities and problems posed by genetic research. "If he's right [that] there's a gene on the X chromosome that affects sexual orientation . . . that's a very, very important find," Bailey says. Combine the highly suggestive twin studies of Bailey and Pillard with Hamer's finding and it seems powerfully likely that genes do indeed influence sexual orientation. We don't know just which genes and we certainly don't know how, but Hamer and other scientists have the tools to begin finding out.

What's most misleading about the premature, high-profile trumpeting of Hamer's findings in the popular media is that it fails to acknowledge how tricky this kind of investigation is, and how many "ifs" are involved. Breathless coverage has provoked some scientists to outright sarcasm.

"Unfortunately, the caveat vanishes completely as word of the latest discovery moves from *Science* to *Newsweek*," says Cornell University's Daryl Bem. "The public can be forgiven for believing that research is but one government grant away from pinpointing the penis-preference gene."

Most scientists who suspect a biological basis for sexual orientation believe it's the gene-to-hormone apparatus that organizes the brain in a way that causes an individual to find the opposite sex—or, sometimes, the same sex—desirable. That seems to be the case with rams, anyway—the only other animals in which scientists have confirmed a lifelong, unwavering interest by males in other males. Before citing some of this research, however, perhaps I ought to confess that when I began learning about it I found the whole notion of gay rams hilarious. For weeks I couldn't even say the words "homosexual sheep" without starting to laugh. My point in telling this is to emphasize that I approached the sheep story with a built-in prejudice, yet found the information impressive anyway.

There's a lot of discussion about same-sex orientation in the natural world. Scientists have reported sexual encounters, mostly male-mounting-male, in quite a few species—at last count 63. But this isn't necessarily orientation, or a lifelong desire for one sex or the other. It appears, rather, to be in almost every case a passing variation: the occasional mount, a flash of interest, even confusion among young males. In most cases, based on observations of captive animals such as monkeys, the animals bounce back into a basically heterosexual life. It's true that female Japanese macaques seem to maintain a lengthy consort relationship with other females, but they still mate with available males.

Among sheep, it's different. There appears to be consistent, unshakable male-male preference among rams, similar to that among regularly practicing gay men. (Sheep do not make a good comparison to lesbians; ewes are consistently opposite-sex oriented.)

Scientists have done everything but dangle gift-wrapped females in front of rams who are male-oriented, without result. In one study, they first isolated eight rams individually for up to three days in the mating season, guaranteeing some pent-up sexual frustration. They then let each ram loose, one at a time, into a corral which contained two ovulating females and two other rams. Six of the rams couldn't reach the females fast enough. Two of the rams apparently couldn't see the ewes, they were in such a hurry to reach another ram. In another experiment, at the U.S. Experimental Sheep Station in Idaho, scientists identified six male-oriented rams in a group of 30. They did so by watching them operate in a herd—repeatedly mounting males and brushing by females.

All the studies show a suggestive influence of testosterone in these male-oriented rams. In the Idaho study, the male-oriented rams were consistently lower in blood-level testosterone than those who preferred females. Under closer scrutiny, it appeared that the testes of these homosexual sheep were not making as much testosterone as those who were opposite-sex oriented. These are preliminary findings, though.

There have also been a host of experiments in which biologists artificially raise and lower testosterone levels in rodents—rats, mice, hamsters—and see an influence on sexual orientation. The same kind of experiment has been done with dogs, which are of interest because they share an ancestry with wolves, a monogamous species, but have grown into a highly promiscuous and, as is well known, sexually excitable group.

Researchers have observed that even given this excitability, female dogs ("bitch" is the correct technical term, but I hate it) consistently want to choose their mates. While in estrus, they will seek out a favored male even though they're often pursued by many. However, when, in one study done with beagles, scientists injected female fetuses with extra testosterone, those females tended to seek out other females instead of males when they reached adulthood (or began to ovulate).

"I do think there's a biology to homosexuality and that hormones play a part, at least in men," Bailey says. "For women, it may be a little more flexible. I don't know why. But my impression, among self-identified lesbians, is they often have sexual arousal to men and choose not to act on it. They choose to be with women. This is one thing that seems truly different in men and women. Women are more likely than men to have feelings toward both sexes. Men are more channeled one way or the other—there's less choice."

In one of his more controversial analyses, evolutionary biologist Donald Symons once suggested that homosexual men represent the human male unfettered by the human female. This argument may seem a little twisted, but it's worth pursuing because it emphasizes the importance of social relationships in human sexual behavior.

Symons' theory was born in the pre-AIDS era of newly liberated homosexuality, those few giddy years when men came airborne out of the closet, when there was an in-your-face flaunting of sexual freedom built around the gay bathhouse and the one-night stand. There was a point at which promiscuity seemed everything. When it crashed, painfully, with the AIDS epidemic, the level of wild coupling proved a major obstacle to epidemiologists trying to figure out what the disease was and where it had begun. They were trying to track it down through men who boasted of hundreds of encounters, often without knowing any of their partners' names.

In a notably calm and persuasive style, Symons has published a biological theory to account for this kind of behavior. This, says Symons, is where biology leads men without women. He argues that heterosexual men would be more than happy to engage in frequent sex with strangers, even to enjoy public baths or quickies in a public restroom, if women would just cooperate. "But women," Symons writes,"are not interested." His point is not that men are—contrary to popular jokes—incapable of commitment. (There's extraordinary lifetime commitment in the gay male community.) Rather, he's arguing that without female influence you will find the sexual extreme of male behavior. And, equally, without male influence, you will find the cozy, nesting extreme of female behavior.

Daryl Bem reinforces this idea. "Suppose you went up to your average, single, heterosexual male and said, there's a new institution called a bathhouse and it's full of young women who want sex, who are wearing only towels. It's safe sex and there's no exploitation. What do you think would happen?"

He asks me this during a California-to-New-York telephone conversation, and I reply, "They'd be standing in line." He laughs. "Right. But if you offered the same thing to women—an institution full of eager young men—most of them would reply that they'd rather know the person they're going to bed with.

"What's interesting about the right wing's objections to homosexual behavior," he continues, "is that what they're fixed on has nothing to do with sexual orientation at all. It has to with men. They're objecting to the way men behave compared to women."

Lesbians don't have bathhouses; they never have. More than any other group (including heterosexual couples), they build long-term relationships and, apparently, engage in sex less often than couples in which at least one of the partners is male.

Bem says that studies exploring the endurance of frequent and enthusiastic sexual relations between long-standing couples find that physical passion thrives best in the homosexual male relationship, then in the heterosexual relationship, and last between lesbians. "There's a joke in the gay community that goes like this," he says:

What does a lesbian bring on her second date?
A U-Haul.
And what does a gay man bring on his second date?
What second date?

"And if you tell this in the gay community," Bem adds, "they will know exactly what you are talking about."

Bailey also comments on the contrast between male homosexual and lesbian couples. He points out, though, that if Symons' promiscuity theory is right, it isn't as simple as it sounds. Consider lesbians. Here you have a woman attracted to a woman—in other words, carrying on as if she possesses the sexual orientation of your average man. Why then don't you get the male sexual attitude? Why isn't sex casual and hot and frequent? The question swings in reverse as well. If homosexual men have, essentially, a female sexual orientation, why aren't they slower and more cautious and choosier? In other words, Dean Hamer may be right—we are tracking two distinct sexualities. Male-homosexual and male-heterosexual are like two variations on one musical theme. Women, straight or lesbian, are playing a different song.

Researchers agree on this: Whatever the formula, the force, or the magnetic pull, it's in place very early. A few hormonal hardliners say sexual orientation locks in by the fifth month of gestation. That seems untestable to me, but their argument is basically that hormones "masculinize" or "feminize" the brain by that point, which includes orientation. This is an extreme position, though. Most believe that sexual orientation slides into position in the first years of childhood, somewhere between the age of two and four years. Either way, the timing suggests that anyone who argues that sexual orientation is a "choice" is clueless. Does a toddler plot his future sexual preference? At that age, my sons were involved in what our preschool called "territory disputes" over toy cars and balls. Translation: They were either biting the child who took away the toy or getting bitten. I can't begin to remember how many notes we have received about applications of ice and TLC.

And yet, as Bem points out, the early slotting of sexual orientation does allow for social and cultural influences on developing sexuality. If you study children and how they play, it becomes clear that, from very early on, we—as parents and as a culture at large—steer them into sex-typical roles. Bem calls this the signature of a gender-polarizing society.

I've read studies that suggest gender awareness can be identified in babies between the ages of six and 12 months. But these "what does baby think" studies always feel to me too flimsy to build a case on. In one experiment, scientists showed infants pictures of men and women, matched with either male or female voices. The infants stared preferentially at the pictures with a properly matched voice rather than those combined with a voice of the opposite sex. There's an argument, and very reputable scientists make it, that such gazing indicates a kind of comfortable recognition of what's right. For me, it's a reach. I'm still struggling with how much you can read into a baby's glance. More and more studies suggest, however, that we have underestimated infant awareness.

I'm more comfortable with accepting gender awareness in talking toddlers. The consensus seems to be that full gender awareness, as in "I am a girl" or "I am a boy," also arrives between the ages of two and three. This sense of gender identity is not synonymous with sexual orientation, although it may eventually play a part in it. Gender identity has to do with one's first awareness of one's own sex and the kind of behaviors that go with that. If a family operates in a very traditional, Beaver Cleaver kind of environment, both awareness and an association with "proper" behavior—boys do trucks, girls do dolls—seem to happen at the early end of this time frame. If a child grows up in a more nontraditional, role-sharing family—"We all do the dishes, Joshua"—studies show that children maintain much more flexible ideas of gender roles until about age six. A series of interviews with children found that three-year-olds say about half their friendships are with the opposite sex. By five years of age, that drops to 20 percent. By seven, almost no boys or girls have, or at least will admit to having, best friends of the opposite sex. They still hang out on the same playground, they may be friendly, but the real buddies tend to be boy-to-boy or girl-to-girl.

Even at preschool ages, boys and girls fall into very different play patterns. Boys tend to gather in larger, competitive groups. They play games that have clear winners and losers and bluster through them, boasting about their skills. Girls, early on, begin gathering in small groups, playing games such as dress-up and house, that don't feature hierarchy or winners. One study of children aged three to four found that they were already resolving conflict in different ways—boys resorting to threats, girls settling for compromise. We tend to reinforce those differences, though. One recent study of children's toys looked at what happened when a child requested a "gender-appropriate" toy, and what happened when the request was for the opposite. It found that if a boy asks his parents for a fighting action figure, they're more likely to buy that gift than if he asks for a Barbie doll. No matter that in both cases, he really wants the toy. The study, which involved almost 300 children, showed that some 70 percent of the time, the boy got the Power Ranger or whatever the boy-type toy was. But his odds of getting the Barbie weren't nearly as good—40 percent at best.

It's amazing how rapidly we divide the world into male and

female, appropriate and inappropriate—not just in toys but in color of clothes, or pitch of voice. Children become ruthless about this, as British scientists have eloquently illustrated. They found that by age eight, children had assigned musical instruments as either boy- or girl-appropriate. Girls played flute, piano, and violin. Boys played drums, trumpets, and guitars. The peer pressure became whiplike. A boy who played the wrong instrument could be written off as a "music geek"; a girl playing trumpet was notably "uncool." When the researchers talked to friends of a flute-playing boy, they rose to his defense, assuring the scientists that the boy was trying to quit.

Bem points out that it seems to be in this particular stage, from six to ten years old, that children lay down some of their most fundamental rules about how to behave. This no cutesy commentary on the laws of hopscotch, either. The great Swiss psychologist Jean Piaget suggested that our basic morality sets at this stage, when social rules take on the inflexibility of gravity itself. Piaget did a terrific marble study, for instance. He found that if you play a game which requires red marbles, and you suggest replacing them with blue, five-year-olds handle this well, ten-year-olds roll with it, but six-to-nine-year-olds, by golly, want the red marbles. And, interestingly, the same applies to gender roles at this age. Ask children if a man can become a nurse, and they'll almost always say yes, except in that six-to-nine period, when it's a woman's role, because those are the rules.

There's a lively debate in science—reinforced by an equally lively debate among parents—about whether the path to such gender separation begins in our biology, whether society merely reinforces a natural split. Many parents today, trying to raise children in a gender-neutral household, have been frustrated by their daughters' insistence on playing house and their sons' refusal to consider joining them. (Stand-up comedians joke that if you give a boy a doll and a girl a gun, she'll tuck the gun into a toy stroller and he'll aim the doll like a revolver.) Countless parents of boys, myself included, who imagined they could influence their children's play choices have quickly learned that they aren't as influential as all that.

My son Marcus, so charming about valentines and hearts, passionately covets toy weaponry. Denied even so much as one lousy plastic pistol by his gun-intolerant mother, he has compensated by building armaments out of everything from clay to kitchen utensils. I watched him charge after the cat, rushing about the house shouting "Shoot him with a toothbrush!" and I found myself mentally throwing up my hands. Back in the 1950s, when boys were expected to be gun-toting rowdies and girls to be doll-carrying little mothers, who spent a minute on this? It just was.

Some of our belief that we are swimming upstream against some deep biological current is surely just modern frustration, but some of it is reality. Bem points out that male human fetuses are much more active prenatally than the females. Since no one, so far, is arguing that society reaches into the womb and instructs the fetus on appropriate gender behavior, clearly there's a strong suggestion here that the more active, rough-and-tumble play style of boys has a biological underpinning. Researchers such as Beverly Fagot, a psychologist at the University of Oregon, have also shown that infant boys react most happily to trains and cars and baby girls prefer dolls very early on. The question is how much to read into this: Are vehicles, as some psychologists suggest, symbols of power, speed, and emotional distance? Or are they merely enticing for their moving parts? Are dolls representative of social relationships, family, and nurturing? Or are we plastering adult symbols all over very tiny people? It is interesting, though, that signs of the division present themselves so early.

It's worth exploring play and gender patterns further. There's some evidence, notably in rats, that sex-typical play is needed for proper brain development. Young male rats who fail to tumble each other around seem to grow up sexually incompetent, perhaps failing to spike high-enough testosterone levels for needed brain organization. That finding, however, has not been clearly repeated in other species.

The hottest theory of play is that it's a practice run at the more serious business of adult life. Play offers a nonharmful way of figuring out how tough an opponent is—I'll push you, you'll push me back. It's a way of measuring the competition, building a relationship—in other words, socialization. Along the way, youngsters learn to work together, absolutely essential for adult cooperation (such as pack-hunting in hyenas).

"Play offers a non-life-threatening way of asserting yourself," says Christine Drea, a postdoctoral researcher at the hyena colony of the University of California, Berkeley. "By playing, you learn skills about competition and aggression." In that context, testosterone provides an edge for males. Research strongly hints that this hormone drives the push-and-shove aspect of play. Back in the late 1970s, Robert Goy, an eminent psychologist at the University of Wisconsin, first documented that young male monkeys consistently played much more roughly than juvenile females. Goy then went on to show that if you manipulate testosterone levels—raising it in females, cutting it off in males—you can reverse those effects, creating sweet little boy monkeys and high-flying girls.

Remember the girls with the Congenital Adrenal Hyperplasia (CAH) defect—the ones whose malfunctioning adrenal glands caused them to be unusually high in circulating testosterone? (See chapter 1, page 33.) Not surprisingly, many of those girls have lived their lives surrounded by gender researchers. The CAH girls, as a group, play pretty tough, and, if offered a choice of toys, gravitate toward trucks and pistols over dolls and houses. Critics point out that you can't clearly separate culture and biology here. CAH girls are visibly masculinized by testosterone, with their penislike clitoris. Their parents, not surprisingly, treat them a little differently than the average girl is treated.

UCLA researcher Melissa Hines, who's done some beautifully detailed studies of CAH biology, added another dimension, which she reported at the 1996 meeting of the American Association for the Advancement of Science. Hines pointed out that how girls play can be greatly influenced by the choice of playmates. If they play with other girls, they tend to fall in with the more stereotypical choices, cheerfully joining in dollhouse play or hopscotch. It's peer pressure—it's the way we respond to an environment, to our wish to belong. It may even be that there's a give-and-take of biology here—that playing with other girls provides a more soothing environment. Of course, in monkeys, just one boy can cause the opposite effect. A young male tumbling into a group of playful girls rapidly turns the whole game rowdy. Hyena play actually offers an interesting insight on these questions. Hyenas represent a counterculture among animals, because the females are revved with testosterone to an unusually high degree. They behave like it, too. Female hyenas as youngsters are far more boisterous than boys: noisier, pushier, rougher. But work done by Joanne Pederson, another researcher at Berkeley's hyena lab, found that if young hyena males play with females, they get rougher; they move toward the wilder female style. In turn, the young hyena females soften their approach slightly if they're hanging out with young males. That pattern parallels human play fairly closely, although, of course, in reverse. If a girl joins boys at play, she runs and tumbles much more than she would with other girls. Boys, though, never relax into girls' games quite so smoothly. Like female hyenas, they seem basically charged with energy, ready to roll. They are less rowdy and less distracted when playing with girls, but they don't alter their play patterns as much. Still, the studies show that "there's a social modulation [from playing together]," Christine Drea says. "And I wouldn't be surprised if some hormonal modulation accompanied that."

The hyenas illustrate another point, as Drea emphasizes. Hyenas, possibly because of high prenatal exposure to androgens, literally are born fighting. Unlike many predators, they are born with fully usable teeth—and they use them. Because of that, hyenas have no need to slowly learn the rules of aggression. They tumble into them immediately. Despite that, these are extremely playful animals. In fact, after that intense period of infant fighting, young hyenas tend to prefer play to all other activities (except, of course, sleeping and eating). That suggests, at least to Drea, that play is important in learning other skills besides how to size up the enemy. Hyenas build their friendships through play; they build bonds that last a lifetime. "They have aggression down pat at birth," Drea says, "so that with play, they learn other skills, such as establishing and maintaining an affiliative bond. In social species, one of the keys to survival is not being isolated—being part of a group."

It's at this point—the way boys and girls play differently, the way they play or don't play together—that the questions of sexual orientation get very interesting. It turns out that if a young boy is very feminine, insisting on dress-up, cookery, dolls, and girls as playmates, that's a strong predictor of adult homosexual practice. Some 75 percent of these superfeminine boys, dubbed "sissy boys" by UCLA psychologist Richard Greene, show a strong same-sex orientation when they grow up. Although they make up only a small part of the homosexual community, the childhood patterns are worth exploring.

As mentioned earlier, childhood can be extraordinarily rigid and unforgiving. As adults, we accept—and even enjoy—people who break stereotypes: the man who stays home with the children, the woman who scales Mt. Everest. But children, remember, can write off another child for playing the wrong musical instrument. Boy rules and girl rules come down very hard. It's slightly easier for girls; tomboy girls are more accepted, by parents and by other children, than doll-toting boys. Although parents may push (and there's plenty of evidence that fathers, in particular, nervously guard against "girl" behaviors in their sons), Bem suggests that it's peer pressure that's most potent here. A child who goes against schoolyard conventions is a child guaranteed a very tough time. That implies a strong biological imperative behind "sissy boy" behavior.

As for parents, they can be gay, straight, or goat fetishists, apparently, without making much of a difference in the sexual preferences of their children. A classic investigation in this regard, by researchers at London's City University, compared 25 children raised by lesbian mothers with 21 raised by single, heterosexual mothers. Did growing up in a household where same-sex relationships were the standard produce children with same-sex orientation? Conservative political agendas certainly predict that would be the case—that "perversion" breeds "perversion." But no. Almost all of the children, no matter what their parents' orientation, grew up heterosexual; two of the 25 raised in lesbian homes later developed lengthy same-sex relationships. Overall, the researchers said, the children from lesbian homes were more relaxed, more experimental, and more willing to dabble in different sexual arrangements, but were always clearly comfortable with their own path.

"I don't think parents do make very much of a difference," Bem says. "And those who try to force sexuality need to try to understand it like this: It may be that you have a boy with strongly feminine traits, and that may mean he's likely to become gay. Your choice is really whether you want him to grow up happy and well-adjusted or miserable and neurotic. If society was very free, open to sexual preferences, then parents might have more say. But our society already has a complete heterosexual agenda underway, and that puts pressure on a child from when he or she is very small. If a child is gender-nonconforming, despite such pressure, the child is likely to become gay. That's still going to freak out some parents—but they shouldn't agonize over it."

Richard Greene's pioneering "sissy boy" work pointed toward a similar conclusion. Of the families he studied, a few were appalled at their sons' "girlish" habits. Fathers (and this pattern shows up in all research) were particularly alarmed. In about 27 percent of the families, the parents were so upset they brought their sons to see psychologists and scheduled them into counseling sessions aimed at discouraging homosexual tendencies. And the result? About three-quarters of those counseled became either homosexual or bisexual, actually a slightly higher percentage than those who received no counseling at all.

Bem and his wife, Sandra, also a psychologist at Cornell, determinedly brought their children up in a gender-neutral household. In the six-to-nine age period, he says, both boy and girl were militant feminists. Their son deliberately wore a bright pink backpack, telling his parents that no one was going to force him into color choices by gender-label. Both children, he adds, grew up to be heterosexual.

Although only a small percentage of adult homosexuals come from the "sissy boy" category, surveys of the community indicate that many men and women remember feeling somehow different, separate, from a very young age. The Kinsey Institute's 1981 San Francisco Study compared 1,000 gay men and lesbians to 500 straight men and women. Sixty-three percent of the gay men reported that as children they had disliked classic boy activities such as baseball and football. Almost half had preferred playing games such as house or hopscotch. That compared with about 10 percent of the straight men. The numbers reversed for lesbians. Sixty-three percent didn't like house and jacks; 81 percent preferred sports as children. Again, girls were less either/or: 61 percent of the straight women also loved sports as children, they just loved playing house too. Mike Bailey's work in Australia, though, found that gender nonconformity in childhood was still the best predictor of whether a girl would grow up a lesbian.

What exactly does it mean, that drive to swim upstream, against the prevailing flow? When your companions—or at least your age-group—build their gender rules out of rock, why would you buck the system? It seems almost heroic for a small child to go against the group. It's partly that heroism, or whatever it is, that has led scientists to believe there is a biology behind this, an internal push stronger than the external push.

"I take issue with the idea of 'born gay,' " says Berkeley's Marc Breedlove. "People may inherit a gene for male-pattern baldness. That doesn't mean they're 'born bald.' Or that everyone with the gene becomes bald at the same time, or to the same degree. Just because you inherit a gene that will cause a certain trait to appear later in life, that doesn't mean you are 'born' with that trait. So, I think the 'born gay' phrase simply spreads confusion by sloppy thinking." But Breedlove also suspects, based partly on Hamer's work and partly on the Bailey and Pillard studies, that there's a genetic component to the behavior. Perhaps it is on that elusive region of the X chromosome found by Hamer. Perhaps, Breedlove suggests, the genetic material at Xq28 increases "the chance that the male will become gay—not born gay—as the newborn, as far as I can see, has no orientation at all. What remains to be discovered is how the gene, or perhaps genes, there affect the probability of homosexuality in adulthood. Does the gene change the probability of interest in dolls, or in modeling the primary care-giver regardless of his/her gender? We have no data to distinguish these from hundreds of other ideas."

Some of the theories of orientation-timing are based on the hypothalamus, which many scientists suspect is where the brain regulates sexual orientation. The hypothalamus is known to play a part in male mating behaviors. If scientists deliberately scar it in the brains of species including cats, dogs, goats, and monkeys, the animals simply lose interest in sex. The hypothalamus is also where Simon LeVay did his well-known comparison of neuron structures in gay and straight men (technically the INAH-3 cells) and found that the brains of gay men looked slightly more feminine. Still, as also noted earlier, if you remove, at least in animals, the INAH neurons, there seems to be no apparent effect on sexual behavior. The hypothalamus, like much of the brain, remains full of promise and mystery.

By tracking patterns of development in the hypothalamus, scientists can raise the question of whether people are actually "born with" these structural differences. As Dutch biologist Dick Swaab points out, there's nothing fixed about the hypothalamus at birth. Consider the Sexually Dimorphic Nucleus (SDN), which is another set of INAH neurons: Boys and girls are born with about a near-identical nucleus. Between the ages of two and four, the nucleus grows rapidly and equally in both sexes. It's only after age four—and usually by age 10—that the cell groups begin to differ by sex. The male cell nucleus becomes larger than the female. What does that mean?

Is the change preprogrammed? There's another testosterone surge right after birth in boys. Swaab has suggested that it may, somehow, order male brains to hang onto more of those nuclei. But we don't know why. Perhaps the difference isn't even relevant to sexual orientation. It may be that something else sets sexual orientation and that that, in turn, directs the assembly of sexually dimorphic nuclei in the hypothalamus. Breedlove notes: "We do not know whether the difference [in the SDN or INAH-1 cells] is a cause or a result of the development of sexual orientation, but most of the public press seems unaware of the latter possibility."

And there's another question that researchers are just beginning to approach: Does the development of sexual orientation only affect sexual preference? Or is it part of a chain effect, linked to certain talents and skills? If a man orients toward men (the way most females do), does his brain strengthen along other so-called female pathways? This is what I think of as a provocative idea; the evidence is suggestive but slight. There are a few studies showing that gay men outperform straight men on verbal tasks and that, conversely, straight men are better at spatial reasoning. Doreen Kimura, a Canadian and a prolific researcher into sex differences, recently published a study comparing gay men, lesbians, and heterosexual males and females on two skill tasks: throwing a ball at a target, something males typically do better, and plugging pegs into small holes, a task that normally favors women. In that study, gay men consistently threw less accurately than straight men; gay women outperformed straight women. On the fine-motor task, however, women simply did better than men regardless of sexual orientation.

In another study, Bailey surveyed straight and gay dancers, exploring the connection between sexual orientation and choice of profession. Seventy percent of the male dancers were gay, and Bailey suspects there's a biology to it somewhere. "I think these occupational stereotypes are very true and they reflect some kind of early interest pattern that has something to do with brain organization. We do well what we like and we like what we do well. If you ask the dancers how they got started, the straight men are likely to say that they got pushed into it by their parents. The gay men are likely to say that they wanted to." There's also a self-reinforcing nature to this. If, as it has, the arts community becomes known as a home to homosexual men, surely homosexual men will be more likely to pursue careers in that accepting environment rather than brave the more hostile culture of, say, an automobile assembly line.

Which came first—the acceptance of gays in creative and expressive professions or the attraction among gays for such careers? It's a chicken-and-egg kind of question, and so, really, is the question of what pushes sexual orientation.

Bem, partly as a challenge to the hormone-lovers, the genetic determinists, has come up with his own provocative theory by mixing biology with culture—two realms of influence that can almost never be satisfactorily separated anyway. As Bem points out, Hamer may have found a gene connected with male homosexual behavior on the long arm of the X chromosome, but it may have nothing to do with sexual orientation. The X chromosome is a busy place; it handles color vision and blood-clotting capabilities. Whatever gene or genes lurk at Xq28, Bem says, could influence personality traits rather than specifically sexual behavior. In fact, he argues, that scenario is far more likely. Some very good science has been done showing that some aspects of personal character, such as shyness, can be strongly influenced by genetics.

Bear in mind, again, that behavioral geneticists do not say that any human behavior is rigidly determined by genes; no one—even the greatest enthusiasts—describe a personality trait as 100 percent genetic. Not even close. The influence of genes on personality traits is usually estimated at about 50 percent (about the same percentage as might be determined by chance). That can indicate a strong predisposition, but it also leaves room for an equally strong environmental effect. There's also some variation in these estimates of the relationship between nature and what we call nurture (as if that part weren't nature, too). Among the current estimates of genetic influence, for instance, extroversion and shyness are at about 50 percent; self-esteem—difficult even to define—seems to be at about 10 percent. Intelligence, or at least performance on carefully calculated IQ tests, is often said to be nearly 70 percent heritable. Bem argues that each of us brings a cluster of such variable genetic influences into the world. These may predispose us to respond to the world differently—and then it responds back.

The traits that have been linked to genes, however, seem remarkably persistent. Jerome Kagan, at Harvard University, has pioneered studies of shy children. In one study of 100 children, followed from age two until age 12, he and his colleagues found that the same girls who were inhibited toddlers were also shy and somewhat fearful teenagers. The boys who were shy two-year-olds appeared more outgoing at 12. But, Kagan said, they were, overall, more emotionally subdued and, in private, confessed to many more fears than children who were notable extroverts. Kagan suggested that our society is less tolerant of shy boys, so that teenagers try to hide it. Most little children haven't learned such deceptive techniques; a small boy who dislikes aggressive push-and-shove may simply choose to play with the girls in his neighborhood or at the day care center.

No one is saying—I am not saying—that playing with girls makes boys gay. Or vice versa. Quite an interesting study of career women shows that many of them played a lot with boys, enjoyed competitive games, and carried that edge right into the business world. But that wasn't a study of sexual orientation. Play choice does appear, however, to be an indicator of how a child regards the world, of what that child wants or needs.

What Bem's theory does is look at "gender nonconformity"—say, small boys who prefer dolls to trucks—in a different light. Consider the tendency to choose a mate of opposing characteristics, to seek out genetic diversity. Bem points out that we tend, as humans, to be drawn toward the exotic as long as that's not too alien. We often marry within our own class, our own educational background, our own race, our own neighborhood. Our biggest leap toward a stranger, someone different from ourselves, is our choice of the opposite sex.

Bem suggests that our sense of the opposite sex as exotic may be cued by our childhoods, by the way we play: boys with boys, girls with girls. Perceiving the opposite sex as strangely, and eventually fascinatingly, different could be part of the way we build sexual orientation. Perhaps it serves as a biological cue, similar to the way that light signals the brain's visual system to begin processing, or hearing language coaxes the brain into building the structures required to use language. If a boy, for instance, grows up in a world of girls and dolls, perhaps girls, and ultimately women, become too familiar to be desirable. Perhaps through this—and a complex of other influences, including genetics—his erotic development centers on his own sex, which has become the exotic one. Bem calls this theory "Exotic Becomes Erotic"—and he hopes it will remind everyone, including scientists, that there is nothing simple about sexual orientation.

I think Bem's idea fits rather neatly with the notion that women are more sexually fluid than men. As cited before, society and parents are much more permissive of young girls playing with either sex, allowing them to be both soccer players and doll lovers. And, interestingly, women as a group are far more tolerant of homosexual behavior than men. One study by the University of Virginia showed that men were far more likely than women to strongly agree with statements like "[Homosexuals] make me sick," or "Laws against homosexuality are necessary to keep down the numbers of gays and lesbians in the population." Colleen Logan, the study's lead researcher, took this right back to childhood: "From day one, particularly with males, all we hear is that homosexual is the worst thing you can be," she said. Our culture is far more intolerant of boys playing like girls than vice versa. "It's okay to be a tomboy. But it's not [okay] to be called a sissy."

If Bem's theory is right, then perhaps girls' flexible early childhood is part of what he describes as a more flexible adult sexuality. Women, more than men, he believes, are free to choose a homosexual lifestyle; they are less strictly locked in by orientation. A woman may be attracted to both sexes; she may enjoy sexual relationships with men and she may still decide that her life would be better with another woman and simply choose to go that route. Bem says: "I have a lot of sympathy with this. In many ways, women are more lovable than men. They tend more to emotional relationships. There are many reasons why women would be more comfortable with other women. If you were a political feminist who'd had a bad experience with men, and who spent most of your time with social or political women, you might decide to center yourself there. And the fluidity of women's sexuality allows them to do so. They may see men they're attracted to, but they decide that they'd rather spend their time with women. And there is validity to this kind of bias. What's the best predictor of violence? Being male. Perhaps one of the simplest explanations for lesbian behavior has to do with happiness."

Let's move, briefly, into the realm of the totally speculative. Scientists such as Bem and Bailey not only describe women as more sexually flexible, they see men as more strictly channeled in orientation—far more likely to only be attracted to other men, or to only be attracted to women. Does this mean that if we encouraged toddler boys to be with girls more, they'd also develop some sexual fluidity? Actually, what we know of biology doesn't seem to support that. So far, if you follow work such as Hamer's, the suggestion is that male and female orientation may develop along separate paths. Hamer worries that Bem puts too much emphasis on what happens in childhood, and not enough on the internal biological drive of sexual orientation. The best analysis should include both, which "fits everything we know," Hamer says. So perhaps women are just going to be more flexible, men more directed. That, in turn, brings us back to childhood play, but from another direction. Perhaps there's something in orientation itself that stimulates play choice. If we force different play patterns, are we, in some unknown way, playing with our own biology? Should we be messing with this stuff at all? Are there deep, biological reasons why boys will be boys—absolutely need to be boys—and girls will be girls?

Speculation is incredibly tempting here, because as Breedlove points out, we really don't know the origins of sexual orientation. And what we do know tends to become overwhelmed by political and moral beliefs. Certain hard-line conservatives insist that sexual orientation involves a deliberate, conscious choice between good and evil, period. They refuse a biology explanation because they believe it will excuse the behavior. Liberals, for a very similar reason, tend to prefer a biological, even deterministic view of sexual orientation. A 1993 Gallup poll found that people who believe that homosexuality is inborn are also people who believe in stronger civil rights protections for gays. They believe that if our society will accept homosexuality as a "no-choice" decision, driven by genes, then that will help erode prejudice against it. Bem points out that surveys indicate that the same people who readily accept a biology of homosexuality argue angrily against biological theories of race and sex differences. In those cases, they believe accepting a biological explanation will reinforce prejudice. He predicts that some of those same people will see his own theory—with its allowance for interaction of biology and behavior—as anti-gay.

This underscores my real problem with the politicalization of sexual orientation research. It's not just that politics can interfere with the investigation. It's that it can blind us to the results. If we can just stand back a little, the study of sexual orientation already tells us eloquently that we cannot separate biology and behavior into tidy compartments any more than we can pull a diamond apart, carat by carat, and still have the beauty of the original stone.

Chapter Six

THE BIG T

Defining Testosterone

Once upon a time—okay, until about 50 years ago—scientists thought the male body contained a kind of biological magic. Specifically, they thought that the testes produced a substance that was all masculine and all-powerful.

We know now they were dreaming of testosterone. And some undoubtedly continue to do so, despite the fact that the hormone has turned out to be neither of those things.

Testosterone's reputation stood highest, in a sense, before people actually knew what it was. Biologists of those earlier days, mostly men, imagined a material of pure wonder. And they went to real extremes to find it. They analyzed gallons of urine. They ground up animal parts. One leading French scientist of the nineteenth century sought to prove the existence and potency of this magical male stuff by injecting himself with pureed dog testes. He insisted that the extract boosted his energy and sex drive and enabled him to pee in a higher arc, a major issue for men, obviously, in contrast to women.

In the 1920s, physicians grafted monkey testes onto aging men, trying to restore their virility. They literally had to turn away volunteers. And still other doctors used ground-up goat testes to treat people troubled by everything from epilepsy to depression. Then, in the '30s, a group of German researchers distilled 25,000 liters of policemen's urine, looking for the primary male hormone. How unfair, after such effort, that they didn't find it. They did, however, find another androgen, a related hormone called androsterone. That discovery reassured biologists that the approach itself was solid, that perseverance would bring success.

The German team next mashed up some 2,000 pounds of bull testicles. While they were still analyzing the results, they were beaten to the grail by another team of European scientists. In a classic example of big not always being better, the steroid hormone testosterone was isolated from the testes of mice (although it's not clear how many pounds it took). Dutch scientists published those results in 1935, along with a description of testosterone's crystalline structure. Later that year, German biologist Adolf Butenandt, who had unsuccessfully pursued the hormone through policemen's urine and bull testicles, reported on the successful synthesis of the hormone. There are many scientific discoveries that go unheralded. This was not one of them.

Physicians hailed it as "medical dynamite," the test-tube birth of "sexual TNT." Four years later, Butenandt received the Nobel Prize for his work in demystifying testosterone, especially for showing the world how to make it in the laboratory. Since that time, researchers have become so comfortable with testosterone, they rarely even call it by its full name. Among scientists, testosterone often is simply referred to as the letter T, capitalized of course. The big T.

The discovery stripped away the hormone's mystique; it was no longer an unknown source of unlimited power. As Butenandt and his Nobel cowinner, Leopold Ruzicka, demonstrated, testosterone is nothing extraordinary in terms of life chemistry. It's a cholesterol derivative. The basic structure is a chunky package of carbon rings, four in all, with some tag-along oxygen and hydrogen gripping the edges. Take the same four rings, tack on a little extra oxygen and hydrogen, and you end up with one of the so-called "female hormones"—progesterone. Alter testosterone's oxygen/hydrogen ratio just slightly and it converts into estradiol, the primary estrogen. All the steroid hormones are stepchildren of cholesterol, an indirect reminder that for all we worry about having too much of that stuff, we can't do without it.

Nailing the shape and structure of testosterone has not, however (as researchers once hoped), revolutionized medicine. Butenandt and Ruzicka won the Nobel Prize for basic science: isolating a hormone and showing how to re-create it. Fifty years of work based on those abilities has not uncovered anything as medically beneficial as, say, the smallpox vaccine or the development of open-heart surgery. That's not to say that the synthetic testosterone products, known popularly as anabolic steroids, are a complete disappointment.

Charles Yesalis, of the Pennsylvania State University, a professor of health policy and an expert in anabolic steroids, describes them as valuable in treating children deficient in human growth hormone. Adding synthetic testosterone helps push the youngsters into a more normal growth curve, adding height and weight. If the disease AIDS causes a person's muscles to wither, synthetic testosterone also can help; the compound induces the body to rebuild tissue and blocks the relentless wasting of the disease.

A host of universities are now experimenting with the idea of using testosterone supplements—taken orally or through a skin patch—to counter some of the effects of aging. They are not trying to create 90-year-old studs. In most men, testosterone tends to go into a steady decline after age 40. In the average man, the hormone's concentration in the blood drops by some 50 percent by the time he reaches age 80. In that same time period, most men lose between 12 and 20 pounds of muscle, 15 percent of bone mass, and nearly two inches of height. Proponents argue that raising testosterone levels back to a "pre-male menopause" high should bring back the strength and energy of youth. "We give eyeglasses to people as they age to maintain visual acuity," says Norm Mazer, of Theratech, a company that developed one of the newest testosterone patches. "Why not give them testosterone to retain muscle strength and prevent osteoporosis?"

What most of us know best about the anabolic steroids, though, is not their bright health promise but their darker effects. They've gained a shady, black-market reputation as athletic cheats. Athletes—both professional and amateur—use them to pack on muscle bulk, build strength, and outdo competitors. Such usage is based on simple logic: one of the most obvious effects of naturally rising testosterone in puberty is the growth of muscle tissue and associated athletic power. Experts say there may be one million steroid abusers in the United States—bodybuilders, runners, football players, and those in a host of other sports from hockey to swimming, high jumping to weight-lifting. From those steroid-pumped athletes has come the notion of 'roid rages, of people possessed by the demon of a testosterone overdose.

Yesalis believes this is a myth. The athletic world has its share of psychos, he argues, regardless of steroid abuse. What we mostly see is the nasty personality of some athletes reinforced by a sports culture that glorifies the physical response. That's not to say that some people couldn't become twitchy on an overdose of steroids. And several studies indicate that heavy users of steroids are more likely than other athletes to also indulge in tobacco, marijuana, cocaine, and alcohol. "The notion of 'roid rages is blown way out of proportion," Yesalis says. "Let's go to any Penn State versus Ann Arbor [Michigan] football game and I'll show you more cases of alcohol-induced rage."

But it's hard to convince people that steroids are not a major source of behavioral problems. After all, they're basically testosterone, right? And today, people don't think of testosterone as the source of sunlike energy. They think of it—based on several decades of well-publicized research—as a chemical source of violence.

Popular magazines now discuss testosterone as the "hormone from hell," the biological driver behind mindless and criminal behavior. Or as one columnist in a woman's magazine asked plaintively when pondering such masculine chemistry: "Are Men Just Born to be Mean?" It's the surly, muscle-bound image that tends to define testosterone these days, exaggerated to cartoon dimensions—a man with biceps bulging like overinflated balloons, biceps bigger than his brain.

Testosterone is seen as a hormone best suited to our ancestors, befitting the internal chemistry of a club-wielding cave dweller: stupid, mean, and male. Researchers who study violence know the jokes, and they also know they're distortions. Randell Alexander, who specializes in the psychology of child abusers, points out that both mothers and fathers beat up their kids. Alexander, at the University of Iowa, is an expert in shaken-baby syndrome. He's often called to testify at the trials of parents who've killed their children. He's seen both sexes charged with murdering kids. Men do it more often, true, he says, but women are more likely to lie about it. "Women want cover stories," he says. "I think it's cultural. We have much higher standards for what a mother should be than what a father should be. [Mothers] have more to lose. But some of my female colleagues like to say that it's really that men aren't smart enough to think of the cover-up. You know, one of those testosterone-poisoning things."

In other words, now that we have a grip on testosterone and its greasy little cholesterol-based body, we as a society seem not to like it very much. But in our own way, I think we're making a mistake much the same as the one the masculinity worshippers made a century ago. We're underestimating both the hormone and ourselves.

After all, testosterone isn't just a male hormone; it's made by both sexes. Men make it mostly in the testes. Women make it primarily in the adrenal gland and the ovaries. Furthermore, the brain can—and does—convert testosterone into estradiol, so that the so-called "male" hormone easily becomes the so-called "female" hormone. It's those kind of findings that are the real payoff from isolating testosterone. They remind us that biology is slippery. They show that there's nothing simple or straightforward about hormones and behavior. In testosterone, scientists find a sophisticated and remarkably flexible bit of living chemistry.

It's that flexibility, though, that makes understanding the link between testosterone and behavior so tricky. Remember that in the developing human fetus, testosterone seems to provide the early signal, at about six weeks, for a male body. That's different, though, from saying the hormone builds a distinctly male brain, even before birth. If that were proven—which it isn't—we wouldn't spend so much time arguing over whether there's a biology behind gender behaviors. We argue over the uncertainties. What's clear is that steroid hormones, testosterone among them, do interact with the nerve cells, neurons, that make up the brain. They bind to those nerve cells and bring them messages.

Scientists discovered that connection when they found that neurons have receptors for steroid hormones. The receptors are essentially a docking port for the body's messengers. They are very specific; a certain receptor will only recognize and accept a certain substance, be it hormone or immune system agent. In the brain, researchers have found neurons' receptors for androgens, such as testosterone, and also for estrogens.

The brain is thus prepared to listen to what testosterone has to say. And that can be a great deal. Once a hormone locks onto a receptor, it can signal for dramatic changes, from inducing cell division to inducing cell death. Most researchers consider this at least indirect evidence that testosterone is capable of altering the brain and, thus, influencing behavior.

Animal studies began hinting at such influence fairly early. A research group at Stanford University, led by neuroscientist Seymour Levine, found that female rats, given testosterone at birth, not only developed penises, but "knew" how to use them. They could be observed thrusting their bodies, as if ejaculating. The reverse was also true. If they blocked testosterone in newborn males, the penis either shrank to a nub or disappeared entirely. And the males showed no trace of usual sexual behaviors. Often, they wandered over to other males and presented themselves in a receptive, female position.

Levine proposed that the hormone—or lack of it—altered the brain's sexual organization. Later, scientists learned to block androgens very specifically. They could, for instance, create a male rat who had normal testosterone exposure everywhere but in the brain. Those studies tend to support Levine's original idea. Such males had all the right equipment, but were clueless as to how to use it. But further research proved once again that hormonal influences are far from straightforward.

One of the most confusing aspects of testosterone is that, in the brain, it sometimes converts into an estrogen. As I mentioned, men and women both make testosterone; the issue is the different ways we use it. But, as animal studies emphasize, the division between testosterone and estradiol can be very blurred. Many mammals, including humans, make an enzyme, aromatase, that helps convert testosterone to estradiol in the brain. The change is remarkably quick, requiring only one chemical reaction. The reverse, by the way, is not true—estradiol does not easily become testosterone.

In newborn rats, the hypothalamus is apparently geared to such chemical alterations. It is rich with aromatase and packed with estradiol receptors. The latest theory is that rats—and this is rats only—use a testosterone-to-estrogen conversion in masculinizing the brain.

To pursue this idea, Marc Breedlove looked at rats with Testicular Feminization Mutation—the genetic defect, also found in humans, which prevents cell receptors from recognizing androgens. Essentially the hormone is there, but invisible to the cells that normally use it. Breedlove found that male rats with TFM never developed a functional penis. But they acted as if they had. They approached females as if they were fully equipped males. He believes that, in this case, the male rats simply skipped the normal testosterone-to-estrogen reaction in their brains. They used straight estradiol (made in the adrenal glands in males) to carry out masculinization of the brain. Apparently the rats circulated just enough of the estrogen to make that happen.

It may be that male brains use estrogens very differently from female brains. There's no indication of the reverse, nothing to suggest that estradiol makes female rats act like males. But the point is, Breedlove emphasizes, that we have arbitrarily labeled things as male or female when they really belong to the whole species. Through evolution, hormones would have been adapted in either sex in whatever way worked best. Conservation is a basic principle of evolution; nothing that works is thrown away. "Natural selection shows no concern for our labels of what is masculine and what is feminine," Breedlove says.

He's careful to add that estradiol appears unusually potent in rats. Humans also convert testosterone to estradiol in the brain, but it doesn't seem to carry the same punch. TFM boys do behave a lot like girls. Obviously, their background estrogens are not driving masculine behaviors in a way comparable to rats. There are some studies that suggest that in humans, the testosterone-to-estradiol conversion may play a role in irritable reactions. But it's important to be cautious with such results; scientists are still trying to understand when the brain converts testosterone—and why.

What we know for sure is that even in adults, testosterone can alter body shape and look. "It's the primary difference in appearance between men and women," Yesalis says. "If a woman receives a lot of testosterone, on a prolonged basis, you will see pronounced changes. The clitoris will get bigger. She will often develop a deeper voice, male pattern hair growth, and male pattern baldness too. Have you ever noticed how many women bodybuilders have that high teased-up hair? That's to cover the bald spots. Give a female enough testosterone and you will virilize her."

He and other experts agree that high doses of anabolic steroids are most potent and produce the most visible effect in young females, whose bodies are still forming. The compounds are least effective in young males, already nearly topped out on testosterone. "Males have to accept the fact that they are what they are and they can't alter it chemically," says Neil Carolan, who counsels young steroid users in Syracuse, New York.

That sharp alteration in female appearance, though, reminds us again that testosterone is not purely male in its effects. If females lacked receptors for testosterone, it wouldn't affect them. It might just float aimlessly and uselessly. Both sexes make testosterone and use it; the really dramatic difference is that men make so much more of it. On average, ten times as much circulates in their blood. As Yesalis suggests, the male system is naturally loaded with testosterone. An anabolic steroid boost, therefore, is less transforming. Extra testosterone can make women look a lot like men—a pretty extreme change—while in men it merely exaggerates what they already have.

Interestingly, scientists suspect that because females are less routinely exposed to high levels of testosterone, they're more sensitive to its effects. Perhaps males get used to the hormone and it takes a lot to push them into response. The kind of testosterone levels that normally circulate in a woman's body would be almost meaningless in a man's body, far below any effect level. In either sex, though, testosterone looks extremely potent. The average man has only about sixty millionths of an ounce of testosterone circulating in his body at any given time. But women have only an average six millionths of an ounce, a bare hormonal quiver in a man's blood.

The background body chemistry—whatever internal settings say how high testosterone will run in our blood—is called baseline. Most researchers believe genetics determine the baseline. Genes, they say, provide the settings, the code for hormonal construction to keep male testosterone levels generally higher than women's. The obvious result is our differing body shapes. But what seems to have captured both the popular and the scientific imagination recently is the idea of a direct link between testosterone and behavior. That's what gives us jokes about hormone-poisoned male brains, about the caveman persona. Although research has not directly connected testosterone to the way men act, it has offered a strong circumstantial case.

Many researchers do see a sometimes unpleasant correlation between testosterone and personality. As the argument goes, if the baseline is high, a man is likely to exhibit a classic T-type attitude toward life—edgy, assertive, in-your-face. There is very good and very provocative science behind that idea. In a curious way, the notion is both complicated and supported by the fact that T levels are rarely stable. They appear to respond positively or negatively to almost every challenge—and not always in the way we might predict.

Think for a minute of an athlete—say a wrestler—in training. Every day he heads for the gym, pumps those muscles. He lifts weights, does push-ups, lifts weights again. You might imagine that such a superjock would be a testosterone-stud, practically oozing the big T out of his ears. The effect of intensive training, though, turns out to be exactly the opposite. In wrestlers who drive themselves to their physical limit, testosterone levels drop so hard they become comparable to the baseline of men who've been castrated—a bare trickle of hormone. Testosterone, in a sense, seems to take overexertion and stress to the point of exhaustion as a signal for retreat. The same phenomenon shows up in new soldiers in boot camp. In that stage of army life where the military traditionally treats new recruits like dirt, trying to wear them down so they can be rebuilt into good soldiers, testosterone plummets like a falling stone. This turns out to be really bad news for anyone trying to get in shape, self-defeating for both army and athlete. As T-levels drop, so does the hormone's powerful influence on building up muscles. An overworked athlete starts losing muscle mass and strength.

It was out of that dilemma that athletes first began turning to anabolic steroids for recovery of lost testosterone and muscle maintenance. The idea, before it went too far, was simply to be able to train like hell and benefit from it rather than losing. Of course, that missed the point that testosterone doesn't stay depressed forever.

Once the soldier marches off to battle, once the wrestler steps onto the competitive mat, testosterone rises again in response to challenge. We see obvious parallels to this in monkeys, who have had their testosterone levels analyzed even more than humans have. If a male monkey sees an attractive and available female, his

T-level spikes up. If another monkey—say a male of higher rank—covets the same female, he may muscle right into the pair, pushing the would-be lover away, displaying the long, wicked canines of a fighting male. At this point, as a fight approaches, the T-level rises in both monkeys—as if driving an internal readiness for battle. After the fight, testosterone stays high or even climbs a little more, maybe 20 percent higher, in the winner. In the loser, it falls by as much as 90 percent.

If you believe that testosterone is linked to behavior, then this pattern actually makes a lot of sense. High testosterone is the hormonal equivalent of cockiness. The winner is still primed to take on all comers; it may suit him to be balanced right on the edge of aggression. The loser? The last thing he wants is some pushy little hormone shoving him into battle again when he's had no time to recover. Low testosterone is the hormonal equivalent of licking one's wounds. It's that very loss mode that seems to operate in exhausted athletes and soldiers, a hormonal signal that they've pushed too far.

"In monkeys, the finding is quite striking," says Kim Wallen, at Emory University's Yerkes Regional Primate Research Center. "Winning a contest increases T in the winner for about twenty-four hours; in the loser, it stays lower for an even longer time. However, if the loser simply sees a sexually receptive female—especially if the winner has moved away—that boosts the loser's testosterone back up. So now you know [what accounts for] the popularity of strip bars: they're where male losers go to get their T back up." The testosterone numbers from monkeys do show that a little sexual voyeurism is a wonderful antidote for fallen testosterone. Watching other monkeys have sex—at some level, comparable to strip bars or porno flicks, I suppose—boosts male monkeys' T-levels up some 400 percent.

Wallen's colleagues Tom Gordon, an associate director of the Yerkes center, and Irving Bernstein, a psychologist at the University of Georgia, have done some remarkable studies of testosterone flexibility working with rhesus macaques. Rhesus macaques maintain a certain status as the bad boys of the monkey world. They are big, hardy monkeys from Asia, quick to react and quick to take offense. Their canine teeth are daggerlike and they use them that way. They fight with passion. They bite, they scratch, and they've been known to yank each others tails so hard that they pull the skin right off. Veterinarians, rather graphically, call this a degloving injury.

Rhesus macaques are also very smart monkeys. In laboratories, they've been taught to play simple computer games. They can use a joystick to shoot down targets. And they are intensely social animals, accustomed to companionship and miserable in isolation. They normally live in a strictly ordered society, dominated by strong-willed and bossy females. As with chimpanzees, rhesus females build a supportive social network, with daughters baby-sitting for mothers and mothers carrying their infants with them during the day. They do not, however, have female-female consorts as Japanese macaques do; rhesus macaques are unfaithfully polygamous and consistently attracted to the opposite sex.

And, as with the equally polygamous chimpanzees, in rhesus macaque society, there's always an alpha, who's climbed atop the male hierarchy. Back in the late 1960s, Yerkes researchers thought that Mr. Alpha would also be Mr. T, brought to power by his supermacho testosterone levels. Gordon and Bernstein began a series of tests, expecting straightforward proof of that idea. The first set of blood draws, in several different groups, seemed to reinforce the idea. The alpha males did have a higher testosterone baseline.

But then the scientists discovered that these levels stayed high only as long as the male was in alpha position. If he lost a fight, they fell dramatically. If the boss monkey was put into an awkward situation, say, caged with a group of strange and unfriendly females, his T levels plummeted. To understand that drop, you first have to realize that a rhesus male, surrounded by a group of females, is almost never in the position of a star surrounded by fans.

Rhesus macaque females are tolerant of males only when they want to be; that's usually in the mating season. They are otherwise easily annoyed by the male presence, and irritated rhesus females are formidable. Because they are so tightly networked, they can easily gang up on a perceived adversary. They fight viciously, too. "Among rhesus macaques, you don't worry about just the males hurting you," Gordon says. "They're all dangerous. Of the severe injuries we see here at the primate center, of those bad enough to treat, many more of them are caused by female aggression than male."

In terms of a survival strategy, then, there's good reason for an alpha male, surrounded by hostile females, to adopt a nonthreatening, good-guy approach. It wouldn't be particularly helpful to strut around in challenge, or to display a testosterone-induced take-all-comers attitude. For monkeys and for us, Gordon points out, it would be a major liability to be trapped by a hormone into an eternal fighting position. We all need the ability to retreat gracefully—or even just quickly. The rhesus macaque studies of testosterone's response to status were so compelling that when Richard Nixon was forced to resign as president in 1972, Gordon got calls from newspaper reporters asking him whether he thought Nixon's T-levels were dropping. (In a noncommittal kind of way, he said he thought it possible.)

Since then, researchers such as Alan Booth of the Pennsylvania State University, Allan Mazur at Syracuse University, James Dabbs at Georgia State University, and Richard Udry at the University of North Carolina at Chapel Hill have compiled a vivid portrait of testosterone rise-and-fall in humans. They've demonstrated again and again that this is a hormone acutely tuned to where—and who—a person is on a given day, week, or year. Each of us may inherit certain baseline T-levels, but they were never meant to lock in place. They fluctuate on a daily cycle and according to daily events; if there was ever a hormone designed to blow-hot-blow-cold, it seems to be testosterone.

Human studies are often done simply by asking people to spit into vials; it turns out that the testosterone level in saliva is an accurate reflection of the hormone's concentration in blood. That discovery, more than a decade ago, caused a major leap in the number of testosterone studies—people being far more willing to spit than to have blood drawn.

Human research suggests that testosterone is responsive not only to physical challenge but also to mental or intellectual competition. First, a few more examples of physical response: By following testosterone in male tennis players, scientists can basically follow the match itself and even, without watching the game, tell who won. (They take saliva samples before the game, during scheduled breaks, and afterwards.) In both competitors, T-levels rise before the match. As the game continues and as the players settle into the match, testosterone begins to drop in both. If a player loses, testosterone falls even harder. In the winner, just as with the monkeys, it rises again—according to theory, preparing him for the next comer. But men don't have to be dashing about a court hitting balls to display that competitive pattern. It's identical in chess players: the same jagged upward streak of hormone before the game; the steady leveling during competition; the sudden jolt of victory and the T-drop of defeat. In fact, it's enough just to watch a sporting event. Researchers found a comparable rise or fall among spectators at a soccer match, depending on whether their chosen team won. (This was an Italy versus Brazil match. The scientists tracked the soccer fans down in bars afterwards. One unhappy Italian became so upset discussing the match, he screamed so loudly that his throat tore. Surveying the bloody saliva sample, he announced to the researcher: "It's my heart that's bleeding, saliva girl.")

There's a definite link, for men, between testosterone and anticipation of conflict. But we don't know this: Does testosterone sharpen that sense of anticipation? Or does anticipation itself cue a rise in testosterone? The tennis and chess competitions clearly raise that question. In both cases, if a winner regarded his victory as mere luck, if he didn't believe that he triumphed through skill, strength, or both, then testosterone did not rise. It didn't drop like a loser, but there was no triumphant leap to it. And this would make sense, too. If you were an early human, battling it out in some ancient forest, and you won, say, because your opponent fell over a rock, you might not want to challenge the rest of the tribe.

This flexibility, let's emphasize, doesn't mean that baseline testosterone is irrelevant. If it was, the baseline male-female difference wouldn't be so dramatic. But it certainly makes the science both more interesting and more complicated. Putting baseline testosterone into perspective is extraordinarily difficult because it flickers up and down so much. If you didn't understand that, and you took a random blood analysis on the day that a tennis player had just lost a championship match, would you get an accurate background measurement? One of the complaints about testosterone studies, in fact, is that they are often based on only a few measurements. That may be too simple a way to accurately measure a complex compound.

We do know, mostly from work in other species, that consistently high T-levels can be linked to chronic impatience and even aggression. Some of the best work on this line comes from John Wingfield, the avian biologist at the University of Washington in Seattle, who has done detailed testosterone comparisons of bird species.

Remember Wingfield's studies of monogamous bird species, such as sparrows, who normally form a dedicated partnership? Those male birds tended to be lower in testosterone overall than their counterparts among polygamous species. The sparrow males' T-levels drop even further as soon as they become parents. Wingfield found he could change the male's attitude toward monogamy by altering testosterone. If he implanted a testosterone pump in a male sparrow, pushing up the bird's T-levels, then all notions of fidelity vanished. The longtime companion was airborne, out of the nest, chasing the available females. The dedicated father, normally so reliable about bringing home dinner, suddenly lost all interest in the squawky little things in the nest. And, by the way, that indifference brought on a precipitous drop in survival of the baby sparrows.

Not surprisingly, then, the males of polygamous species—flashy cardinals, pushy blue jays—tend to have higher baseline T-levels.

But what is especially interesting is that those T-levels seem relatively constant, less responsive to the vagaries of life. The polygamous birds stayed cranked, more wired, more aggressive. In fact, knowing the connection between hormone and behavior, anyone who observed blue jays and sparrows could make a strong guess as to which species was running on high-octane testosterone. Wingfield wonders whether there's an evolutionary feedback loop wherein the responsive father bird develops a flexible hormonal chemistry, and more flexibility in his response to the world around him allows him to be a more responsive father.

He theorizes that parental care played a role in natural selection: "If expression of parental care is high, then T must be held in check." A male bird that takes no role in raising young can be more consistently aggressive. But a dad that is crucial to the survival of his offspring can't be so edgy that he's dangerous to them; neither can he afford to risk them becoming fatherless chicks because he constantly rushes off to fight. What he needs is a system that will tend toward calm, except in times of danger or instability. Then a jolt of testosterone could be very useful. Say that an invading male is entering the territory, one who might even attack the nest. "Then the ability to increase T secretion to combat the intruder is crucial," Wingfield notes. The same quick response might be helpful if the intruder seemed interested in seduction. "T secretion appears to be most flexible in species in which males show parental care. How this relates to mammals and humans, I'm not sure. This needs careful thought, because mammal and bird parental care systems are so different."

If you were to carry over Wingfield's bird observations into the realm of human behavior, it would suggest that high-testosterone men should be, basically, birds-on-the-wing, prone to promiscuity, indifferent to their children, and quick to perceive insult and take offense. Wingfield cautions against approaching this too literally or assuming that we understand how the hormone works: "Testosterone is certainly very well studied, perhaps the best-studied steroid hormone. However, it is also tainted by urban legend and popular literature, which can suggest facts that are actually untrue, such as the idea of testosterone-poisoning." It may be, he says, that testosterone's real influence is in the realm of sexual behavior, including mating and establishing pair bonds. Those behaviors might result in aggression—for instance, guarding a mate—but the influence would be very indirect.

It's worth noting, I think, that humans do have a highly responsive testosterone system, more like the monogamous birds' than like that of the polygamous ones. And if you look at marriage and testosterone directly, you do find some interesting analogies to birds. In a 1993 study, Alan Booth and James Dabbs looked at T-levels across a range of relationships, from dating to marriage to divorce. High-testosterone men seemed more likely to make a mess of marriage. They were less likely to marry at all and if they did, they were far more likely to divorce, often because they were unfaithful or physically abusive.

But it also seemed that testosterone levels didn't merely predict the relationship—they responded to it. Single men tended to be higher in testosterone than married men. This makes sense if you consider a single man somewhat comparable to a mate-seeking male in other species—out there in the competitive fray. Along that line, you might predict the next finding, this time from Allan Mazur: once married, once settled into a comfortable and stable relationship, men's T-levels dropped. (Again, as mentioned earlier, there's some indication that men's testosterone levels, as in birds and mice, drop again as they become parents.) Also, in our species, instability changes that. Booth and Dabbs found that testosterone levels rose steadily if the marriage began to falter, and rose higher if it moved to anger and argument. T-levels also rose if the marriage broke up and the man moved back into competitive dating. As Wingfield does with birds, we can speculate here about the existence of a feedback loop: Do men with high T-levels tend to have bad marriages? Or does a bad marriage push testosterone levels up? And if so, do those higher levels produce edgier behaviors in a husband, making the marriage worse?

This leads us to the chicken-or-egg dilemma of linking testosterone to violence. Dabbs points out, for instance, that despite more than 20 years of searching, no one has been able to identify criminals and noncriminals by their testosterone levels. You can find high testosterone levels among prisoners, but you can also find them in successful and law-abiding men. Given that, Dabbs has tried narrowing the focus, comparing prisoner to prisoner. Among criminals, there appears to be an association between higher testosterone levels and more vicious crimes. Along with his Georgia colleagues, Dabbs recently surveyed almost 700 inmates in a state prison. They found that car thieves and burglars tended to have lower T-levels than armed robbers and killers. Even more consistent, though, was that higher-testosterone inmates, no matter what their crime, tended to be those in constant trouble, the in-prison rule breakers. Most particularly, they seemed to enjoy rule-breaking if it involved confrontation. They liked facing someone else down.

Dabbs began to suspect that the hormone could be better linked to attitude than to violent behavior. Men with high circulating testosterone simply tended to be more edgy, more in-your-face, more take-no-shit-from-nobody. It's a theory only, but he thinks it could help explain the difference in violent behavior among criminals, and perhaps some of the gender difference in aggression as well. Again, this would be indirect. If testosterone helps make a person more quick to challenge or snap back, then those reactions could produce violent results. In an evolutionary context, the hormone might have become part of a competitive response, even an ingredient in success. It might be a real advantage for a male seeking alpha position to be primed for confrontation—as long as he could also drop back with failure.

Robert Prentky, who studies violent rape, arrived at similar conclusions. Prentky, based at the private Joseph J. Peters Institute in Philadelphia, spent some time in a frustrating search for a connection between high testosterone and tendency to rape. He didn't find it. He found differences within the community of sexual abusers, but it wasn't clear what they meant. For instance, rapists of adult women tended to have higher testosterone levels than child molesters. There was a slight connection between violent rape and higher testosterone. The highest reading came from the only rapist who also murdered his victim; his testosterone levels were nearly twice as high as the average rapist and about two and a half times higher than any of the child molesters. Prentky noted, as Dabbs had, that none of the readings were out of the normal range of testosterone baseline in the overall male population. Instead, Prentky said the most consistent finding was that higher T-levels correlated with anger and with a desire to strike out physically in response to it.

One point stressed by scientists who study testosterone is that if the hormone pushes an aggressive response, that doesn't inevitably mean violence. It doesn't have to result in criminal behavior or murderous anger. There are many ways to channel a pushy and competitive personality. Aggressive hockey players tend to be high in testosterone; so do virtuoso criminal lawyers. Ministers, interestingly, tend to have fairly low levels. One carefully crafted, but very small, study looked at testosterone levels among four young physicians on a boat together during a two-week holiday cruise. Each man brought a woman as a companion. The researchers had the doctors provide a series of saliva samples from the trip. After making the testosterone measurements, the scientists interviewed the women, asking them to rate the men according to how much each physician wanted to be dominant—the alpha male—and how aggressively he pursued it. Although they were not told the results, the women's answers accurately predicted the man with the highest testosterone levels. In this case, the link between high testosterone and the quest for dominance looks fairly real.

Obviously, there are abundant theories on the connection between testosterone and aggressive behavior, meaning that no one has the answer. "It plays a role, but we don't know how much of one," Dabbs says. There's evidence, again from monkeys, that the hormone responds not only to challenge but to the individual's very specific social status at any given time.

Thomas Insel, the director of the Yerkes primate center, and a colleague from the National Institute of Mental Health, James Winslow, once explored some of those complexities in squirrel monkeys, a species of tiny South American tree dwellers. These animals were housed in a laboratory, caged in pairs. The male-male pairs quickly established a dominant-submissive relationship. Overall, perhaps because of winning boss status, the dominant males were higher in testosterone. And, when both monkeys were placed with a familiar female, the high-status males were much more sexually aggressive.

Insel and Winslow did not try to alter testosterone levels. They were interested, rather, in whether testosterone interacted with another hormone, oxytocin, the one so often linked to bonding and social connection. Although oxytocin's role in males is still somewhat mysterious, Insel and Winslow decided to explore whether, if they injected it into the squirrel monkeys, they could see a behavior change. They also wanted to find out if that behavior would be different depending on the monkey's status.

It turned out that status was the key. The scientists found that after oxytocin injections, the dominant monkeys became more aggressive and more sexually assertive. That didn't happen in the subordinate monkeys. Their speculation was that while oxytocin, as expected, stimulated the males' interest in females, it was status that seemed important in how that interest was demonstrated. Dominant males became much more insistent about it; subordinate males remained tentative.

Insel and Winslow speculated that testosterone might be part of the chemistry behind those differing reactions. Studies in rodents had shown that steroid hormones such as testosterone can induce the brain to create oxytocin receptors. (The same is true with another hormone associated with bonding, vasopressin. If testosterone is suppressed in monogamous prairie voles, the males tend to lose interest in relationships. Scientists think that this is because the brain receptors for vasopressin don't get made in sufficient numbers.)

Perhaps, then, a squirrel monkey with high testosterone levels has an extra-large number of oxytocin receptors. And perhaps, again, a charged-up brain response stimulates the animal to act like a bossy and confident alpha male. The subordinate animal doesn't get that extra kick. And perhaps he never will. Perhaps being the low monkey in the relationship keeps his testosterone levels down, his oxytocin receptors confined. It's purely speculative, but it emphasizes the idea that testosterone does not act in isolation. It responds to status, desire, competitiveness. We actually don't know all the conditions to which it might respond. We know just that testosterone does rise and fall to the situation. The hormone's very unpredictability can produce all kinds of behavioral fallout.

It's relatively easy, though, to talk about testosterone, dominance, and aggression in monkeys and rodents. The biology of violence in humans, though, is extraordinarily difficult, partly because it tends to be judged by political standards as well as scientific ones. Behavioral geneticist Greg Carey, of the University of Colorado in Boulder, points out that we already have a simple genetic screen of an unborn fetus that measures risk of violent behavior as an adult. Doctors simply have to check for the Y chromosome; men are seven times as likely as women to be arrested for a violent crime. And so, does that fairly label all boys as potential thugs? The same dilemma exists when you focus on a specific biological factor, such as testosterone. The high-hormone sex (male) is more aggressive, but an individual high-T man may be a decent and law-abiding member of society. An individual woman may be a murderer. Neither the chromosomes nor the hormones provide an obvious explanation for the difference.

Evolutionary psychology offers some reasonable speculation as to why males, in general, seem so much more aggressive than females. The theory is not so different from those which explain risk-taking or emotional differences. It hinges on the fact that we descend from a mating system in which males must compete hard in order to become fathers, and in which females work hard to raise and support the young. That male reality demanded aggression and rules with which to contain it—hierarchy, competition, dominance. The testosterone drive is part of that, a source or a result. While females also had to compete, sometimes for mates and sometimes for food, their primary goals were social support, child care, and child protection. Their reality didn't stand on quite so sharp an edge.

What I'm describing, of course, is a polygamous social system. In monogamy, life is different. John Wingfield emphasizes that monogamous bird species not only tend to be monomorphic (the sexes look the same), but females usually have the same testosterone levels as males. Wingfield suspects that stems partly from the fact that they share in defense duties, such as protecting the property boundaries. This seems to me to be once again a reminder of our ambiguously monogamous nature. Men have the flexible T-levels that characterize monogamy and parenting. But the fact that testosterone varies so widely between men and women hints, once again, that we spring from a polygamous past.

It also needs to be emphasized that testosterone is only one part of the biology-and-aggression question. It interacts with other hormones in a complicated balance of power. There are certainly other strong biological influences on aggressive response. As an example, there's the brain's acutely sensitive use of the neurotransmitters noradrenaline and serotonin in regulating the flight-or-fight response. This combination is not only potent in all of us, male and female, it's extremely responsive, like testosterone, to the pressures of the world around us.

Noradrenaline and serotonin are an incredible balancing act. Serotonin is essentially the soft voice of reason: calming, controlling impulses, regulating against aggression. In an argument with a child, however irritating (and they can be extremely so), most of us do not resort to fists. That is, in part, due to the pull-back effect of serotonin, the introduction of a cooling system. Noradrenaline, by contrast, is the primal scream, all reaction and action. Have you ever been in a near-miss car accident and felt your heart speed up, and had your foot on the brake before you consciously thought of doing so? That's noradrenaline, blowing spark into fire.

The two neurotransmitters ebb and flow in the brain, in rough synchrony, so that a person with higher serotonin will tend to be lower in noradrenaline, and vice versa. Again, there appears to be a genetic baseline for these neurotransmitters. Women, in general, are about 30 percent higher in serotonin than men, reinforcing the notion that females are less reliant on a fight response. Unusually high noradrenaline has been found in a series of violent men, such as Finnish men imprisoned for assaults, and U.S. Marines discharged for abusive behavior. Very high levels of noradrenaline are also found in people who attempt suicide in the bloodiest possible ways.

However, neuroscientist Bruce Perry, of the Baylor School of Medicine in Houston, points out that—as with testosterone—the balance of these neurotransmitters is acutely responsive. Contact with kindness can be physically calming; it can raise serotonin levels. University of Minnesota studies show that as jittery infants, born with high noradrenaline levels, relax into the security of a loving home, their noradrenaline levels ease down too. By contrast, Perry says, vicious and violent experience sends noradrenaline screaming upward. He's found this pattern particularly in stressed-out children of the inner cities. In an environment where fast reactions can save your life, the slow-down, think-about-it influence of serotonin can be a foolish response.

And studies of inner-city residents, interestingly, find a very comparable pattern with testosterone. Allan Mazur and Alan Booth did a detailed analysis in which they found that young men living in dangerous urban environments tended to maintain—almost to an individual—unusually high testosterone levels. "After a while, we began to suspect that what we were seeing was pure response to stress," Mazur said. "These men were living in an atmosphere where they were constantly being challenged. It was survival. I don't think it's surprising to find that they had elevated testosterone."

As my own appreciation for the complexity of testosterone began to grow, I began to wonder how the hormone affected women.

After all, we women make testosterone too. Is a woman with a higher baseline of testosterone more aggressive or competitive? If a woman chooses a challenging profession—my own field of journalism, for instance—does that drive her biology into a more combative state?

I put this question to Kim Wallen, at Emory. It turned out that he doesn't think testosterone operates in women as it does in men. I still liked what he had to say: "What I find intriguing," he replied, "is the possibility that a woman's testosterone is not affected by environment in the same way as a male's testosterone. This would mean that she would never have the increase from winning, but also never have the decrease from losing. In other words, she would be buffered from the fluctuations that males have to deal with due to the [extreme responsiveness] of their testosterone system."

Wallen continued: "This is actually a more reasonable scenario since it seems to follow the sort of differential selection that has acted on males and females. The male pattern of 'live fast and die young' can take advantage of the momentary burst of testosterone to push a momentary advantage. Females, on the other hand—I'm talking about primate females—have a vested interest in long-term stability. This comes from nurturing their slow-growing offspring to independence. No short-term bursts of testosterone are going to facilitate that. Thus, I would think less about how women might be changed by a competitive environment and more about how they might change such an environment by a less-mercurial (my wife's favorite description of me, that and a Ping-Pong ball) response."

The idea that women would be naturally steadier in a high-stakes situation also appeals to me (although I will also confess that I'm the Ping-Pong ball in my marriage). Wallen is correct in that if you consider our evolutionary history, at least what we understand of it, there are good reasons why hormonal influences might be different. That doesn't mean that testosterone might not also be influential in the way women behave. Assuming that women manufacture the hormone for some reason, then it's still worth exploring what little we do know. The preliminary results are downright provocative.

There's some suggestion that if you pump women full of synthetic testosterone—as in steroid-sculpted bodybuilders—they develop a progressive nastiness. A 1994 study by Dutch scientists found that female athletes who boosted their training with the use of anabolic steroids became rapidly angrier and more easily irritated by the people around them. But that did not translate to aggressive behavior. They didn't start shoving people around. They just became grouchy and obnoxious.

My favorite woman-on-testosterone story comes from James Dabbs, of Georgia State, who studied testosterone in female prisoners. This is a single case, meaning that it's statistically meaningless. But it's a great, if gruesome, story. The woman was in jail, indirectly because her husband had been on the telephone. That is, he had been talking on the phone for a long time and that annoyed her. She ripped the phone out of his hand and started beating him on the head with it. When he fled and locked himself in a bathroom, she got their rifle from under the bed. She beat a hole through the door with the gun, fatally shot him through that hole, and then smashed the gun apart on the floor.

The cautionary point is that while the story certainly reminds us that women can be violent—in this case, spectacularly so—it's not proven that high testosterone turned this woman into a gun-smashing lunatic. There's more to that story, a lot more, than a single hormone. If the correlation were that simple, after all, you'd expect all those steroid-pumped female athletes to be literally murderous on the track. Since we can't argue that high-testosterone men are exceptionally dangerous, we can hardly make that case for women.

In the past couple of decades, there have been a few studies linking testosterone with women's behavior. The first notable one was published in 1980. It compared two androgens, testosterone and androstenedione (a chemical that sets up the production of testosterone), in women across a spectrum of careers and lifestyles, from executives to homemakers. The researchers found, in general, that women who occupied "traditional" roles—housewife, clerical worker, salesclerk—had lower androgen levels than those in professional careers or pursuing such careers through college. One widely cited interpretation of the study was that, in this case, the androgens might serve as status indicators. That is, women were more stressed as homemakers or in low-level jobs. Such subordinate positions were depressing their androgen production, something comparable to the beaten-down soldiers. The more "successful" women pursuing traditionally male occupations, apparently, felt like winners and had higher androgen levels.

Of course, there are other ways to read the study. Speaking as a working parent, I'm not convinced that mastering office politics confers greater status—or offers greater rewards—than raising children. But parenting does not require the competitive oneupmanship that corporate life often does. So the androgen results make me wonder, instead, about competitive natures. Might the women who pursued high-end professional careers bring with them a higher-testosterone background?

Richard Udry, from North Carolina, investigated that kind of question in a study titled "Androgen Effects on Women's Gendered Behavior." The study was built on a resource of unusual depth: The Kaiser-Permanente health maintenance organization, based in northern California, has been conducting a study of child development for more than 30 years. As part of that study, Kaiser researchers collected blood samples from pregnant women in the 1960s and froze them. They then continued to follow the health of the women's children.

When Udry learned of the program, he saw a rare opportunity to look at testosterone in two generations of women. He could look at the mothers, going back to their pregnancies through the frozen blood samples. He could test the now-adult daughters. He and his colleagues traced some 350 daughters of the women in the Kaiser study. The women agreed both to give blood samples and to extensive interviews.

Superficially, the results of the study closely resembled the 1980 findings: Women who choose a professional career tended to be higher in testosterone than those who stayed at home. Udry extended the analysis further, though. His survey team interviewed all the daughters in depth, asking a broad range of behavior and lifestyle questions. The questionnaires ranked the daughters on a scale of masculine- to feminine-style behavior. It's important to note that these were traditional American definitions; they weren't meant to define who or what men and women should be. The scientists based them on the country's culture during most of the century. For example, ranking marriage as very important put women high on the feminine end of the scale. So did having children; the larger the family the stronger the score. Putting career emphatically first was masculine, as was refusal to do all the housework and a tendency to thank God for child care. The scientists later correlated the answers with the testosterone readings.

High-testosterone women tended to be less likely to have children, less enthralled by the whole notion of parenting. Lower-testosterone women usually had a great deal more interest in children and in dressing up. They liked makeup; they liked jewelry. They liked cooking better than the high-testosterone women did. They enjoyed interior decorating more. Overall, the high-T daughters tended to be children of high-T mothers. That raises a familiar question of feedback between biology and behavior. Is this a genetic predisposition of some kind, passed along from mother to daughter? Or did the high-T mothers treat their daughters differently—pushing them toward more competitive pursuits—than the low-T mothers? And did the extremely challenging jobs bring up the daughter's testosterone levels?

Although his study rated behaviors on a masculine (career-track) and feminine (makeup and cookies) scale, Udry emphasized that what he was looking at is a normal spectrum of female behavior. "By masculine, we're not saying that these women would be mistaken for men," he says. "These are normal women, the kind you and I know. Their behavior is within the normal range for healthy women. Look among your own friends and you'll find a package of behaviors. Some have more feminine habits, others veer in other directions."

I have to admit that when I started reading Udry's study, I fell right into mentally testing myself. Was I high testosterone or low? There are weekends when I love baking cookies and taking the kids to the park. And there are weekends when I can't wait for Monday so that I can escape to the office, talk to grown-ups, even sort my way through a difficult story about the genetics of yeast. Did any of that represent an ebb and flow of testosterone, I wondered. Or was I getting too weird about this topic? The funny thing was that every female friend I talked to about Udry's study responded by starting to make a similar list: "Well, I wear makeup but I play on the company softball team. . . ." And Udry, while admitting such analyses can be fun, says they are really no more than a game at this point. Although such testosterone-behavior correlations are fascinating, scientists don't know what they mean: "We're working in the dark and we haven't a clue," Udry insists. On the other hand, he says, the findings are more than pure coincidence.

Alan Booth, of Penn State, has been exploring the connection between hormones and behavior for years. He recently completed a study which highlighted yet another steroid hormone, cortisol, which is made in the adrenal glands. Booth has been concentrating on cortisol recently because of the possibility that, more than testosterone, it drives a flexible response in women.

Cortisol is also acutely attuned to stressful situations. It rises not only in a dangerous situation but in a strange one. Researchers have measured its rise in young children, left for the first time at a child care center. In women or men, a new job, a new city, even going to a party where you don't know anyone can bring up internal cortisol levels.

Again, everyone inherits some baseline quantity of the hormone. The cortisol background may also influence behavior. Jerome Kagan, at Harvard, has found that shy children tend to be generally higher in cortisol than more extroverted youngsters. The high-end concentration creates greater muscle tension, which, Kagan points out, tends to make faces less expressive. An outgoing child tends to have lower cortisol levels, a more exuberant personality, and a face vivid in its reactions to others.

This is a subtle difference. These are all cortisol levels within the bounds of natural variation. We're not talking about children with mask-stiff faces, merely those with slightly less responsive expressions. Still, I could argue that the impact on facial expression could affect a child's life profoundly. We all like someone who openly responds to us, whose affection or humor we can read easily. It's not a surprise that outwardly giving children find it easier to make friends than withdrawn ones. And a sense of rejection may cause a timid child to retreat further. In this sense, cortisol may have an indirect, but surprisingly potent, influence on friendship. But it appears to have other equally strong effects.

In a study of women playing college basketball, Booth tracked both testosterone and cortisol levels in their blood. The women did not spike testosterone; it was cortisol that responded to stress, to the competitive situation. And it went down, especially in the most competitive women. Booth asked women to assess their performance. The more confident a woman felt and the better she judged her skills on the court, the lower the cortisol level.

I've tended to think of competitiveness as being all about intensity and tension. But the cortisol response suggests a yet another dimension—that good competitors get cool, and relax with the challenge. That's what Booth thinks he sees in women basketball players. The good ones get calm and focused. Competitive male athletes do the same thing; a comparable study of male tennis players found that the better ones had consistently lower cortisol levels and were more relaxed before the game.

Booth believes that for women, the drop in cortisol allows them to approach competitive situations more like a traditional male—focused on the win. "My notion is that for many women, competition sort of runs up against what they'd naturally like to do," Booth says. "It requires some special adaptation to get them into a competitive mode, to put aside a tendency to be nurturing, to look at others as the enemy. I'm persuaded that most of the difference goes back to evolution, to males battling throughout human history. There's just a tremendous gender difference in willingness to get into combative situations. That's not just in a physical sense, but in other confrontations too, say, at the office."

But there are unanswered questions. His study looked at self-assessment and cortisol, rather than linking the hormone to pure performance. There's a possibility that cortisol correlates with optimism or pessimism about onesself. And there's a real possibility that the underlying confidence or lack of it affects performance. Booth admits that many open questions remain. He plans to look at cortisol and testosterone again. His next study is a long-term one, involving some 400 families. He and his colleagues plan to study both hormones in mothers, fathers, daughters, and sons. One of the pending questions is whether, in a 1990s culture, the girls' hormone levels began to veer toward more masculine levels. If so, would testosterone steadily rise in women over the long term? Would cortisol drop—and would they behave increasingly like men?

"We want to see whether a consistent pattern of parents pushing girls toward traditionally male occupations results in a change," he says. "This is more of an idea than anything else. But in a few years, I hope we'll know more about the reciprocal relationship between hormones and behavior."

In the current study, Booth is particularly interested in whether behavior might influence hormone levels, and whether the resulting hormone levels would influence behavior. And then . . . you can turn the feedback loop forever, it seems. But that variability, that interconnectedness of influence—inner and outer, back and forth—is something we tend to miss when we think of hormones and what they mean. There's a lot of quick political reaction to theories about a cause-and-effect role for testosterone in competition and aggression. Feminists become understandably annoyed by the oversimplified, back-to-the-kitchen notion that women don't have the hormonal underpinnings for competition. And plenty of men—masculinists, if you like—are equally annoyed at being dismissed as a bunch of naturally bad-tempered apes.

Annoyances aside, I think the connections between body chemistry and behavior deserve our attention. Dismissing these theories without giving them fair hearing won't make them go away. I don't think it's insulting to either sex to suggest that hormones influence our reactions, especially if the suggestion is also that how we behave—even how we choose to behave—can influence the hormones. There may be evolutionary influences on who gets how much testosterone and what it does. Still, we're an evolving species; we're not immovably stuck in whatever ancient paths our ancestors trod. Overall, Greg Carey is right: men are more violent than women. If we could explore all the whys—without denying the parts we don't like or agree with—then it might be easier to do something about it. As testosterone's variable nature should remind us, people can change.

Chapter Seven

THE CYCLE GAME

Biology and Estrogens

In an unfortunate kind of gender equality, our culture also stereotypes estrogens.

If testosterone is caveman-with-club, estradiol is woman-from-padded-cave. You know the story: a husband comes home and his smiling wife has made him a birthday dinner. He sits happily down to pot roast and reaches for the salt shaker. What?!? He doesn't like the way she's seasoned the meat! Why did she cook dinner anyway? She marches into the kitchen, pulls out the triple-chocolate layer cake, and dumps it on his head. Oh well, you never know what to expect from a woman in midcycle.

The difference is that the flaky image of woman has been around a long time, maybe as long as human culture, while the image of the hormonally unbalanced man is new. Testosterone just got its nasty image. Estrogens such as estradiol, female chemistry in general, and women's body parts have traditionally been distrusted. Remember that the name "androgen" comes from the Greek word for man; "estrogen" from the Greek word for frenzy. And don't forget that our term "hysteria" descends from the Greek for "suffering in the womb."

In the days when physicians were injecting themselves with ground testes to increase their virility, they were also removing ovaries from women to decrease their moodiness. The doctors of the time described ovaries as a source of female weakness. In 1889, the U.S. Surgeon General's office reported that 51 percent of all operations to remove ovaries were necessary to "repair" mental disorders. American mental institutions of the time kept gynecologists on staff so that they could quickly perform the surgery to calm down the "hysterics." With the turn of the century, that number dropped, but only slightly. In 1907, the government recorded that 42 percent of ovariectomies were used to control behavior. During an earlier period, in the mid-to-late 1800s, doctors in England tried slicing out the clitoris as well as the ovaries. An 1866 letter from a doctor, in a British journal, predicted that ". . . it will be soon somewhat rare to meet with a woman whose sexual organs are entire."

And what makes this so particularly infuriating, in retrospect, is that all of today's science indicates that estrogens—estradiol being the major one—are remarkably good for women. They appear to greatly protect health and even mental well-being, strengthening the body from bones to brain. Most recent research questions assumptions about the links between estrogens and undesirable behavior: Do female hormones bring on depression? Is premenstrual syndrome (PMS) a myth? In hindsight, those misguided surgeons of the past, who removed countless ovaries, probably contributed to the female reputation of fragility by doing so.

That's not to hoist estrogens up to testosterone's erstwhile status—as a source of perfection. But no one today proposes removing ovaries or the clitoris to improve a woman's outlook on life or ability to contribute to society. (In a role reversal of sorts, there's instead an ongoing debate over the merits of castration to control male criminal behavior; for example, California passed a law in 1996 that keeps sex offenders with two convictions in jail indefinitely, unless they agree to either surgical or chemical castration.)

There's no doubt that female bodies (and, in less clearly understood ways, male bodies) rely on estrogens to keep the machinery running. More than 300 different receptors for the hormones have been found, in cells from brain to skin. In both sexes, estrogens appear more directly necessary to staying alive than androgens. There are common, obviously survivable defects involving testosterone and its steroid relatives: cells fail to make androgens, or cells lack sensitivity to them (as in the TFM syndrome). There's the case of CAH girls, who, through a defect of the adrenal gland, get too much androgen exposure. There's no equally common defect of estrogens, and that strongly suggests that most such failures are lethal at a very early stage of development. I emphasize "most" because scientists have artificially created mice with defective estrogen receptors. They're sterile—both male and female—but not stillborn. There may be other survivable estrogen deficiencies, as yet undiscovered. The medical literature reports one known case of a man whose cells lacked proper receptors for estradiol due to a genetic defect. He had slightly deformed bones and, like the mice, he was also sterile. Researchers are still awaiting a second living case of this syndrome. The rareness of defects like this seems to underscore that the estrogens are essential.

Like testosterone, they are clearly part of a shared male and female biology. They may be more complex and more influential in women, but they are not—as people long argued—exclusively "female" hormones. There are actually three estrogens—estrone, estradiol, and estriol—forming a closely related family. Men make estradiol (mostly by enzyme conversion of testosterone, not only in the brain but in other cells, including fat, skin, and blood). Women make all three estrogens. Each is concentrated in a different part of the body and has its own place in the female life cycle.

The estrogens vary only slightly chemically; each has the same basic carbon-ring structure but a different number of oxygen and hydrogen molecules attached. Estradiol, made mainly in the ovaries, is the primary estrogen. It starts slowly rising in girls at about age eight. It continues to rise until puberty, starting about age 11, when it helps induce the menstrual cycle. It remains dominant until menopause, with one exception—pregnancy. During pregnancy, estradiol production shuts down. In its place, the placenta makes estriol, which stays high until a baby is born. At that point, estradiol kicks back in.

As you might expect, then, estrone is the primary estrogen of menopause. As the ovaries shut down production of estradiol with the onset of menopause, the fat cells start making estrone (although, as with men, women's skin and blood cells can also make estrogens). The production of estrone, though, is far less intense. Physicians think that it's the plummet of estrogen that brings on many of the health problems that plague women after the age of 50.

The estrogens tend to work in concert with another set of hormones, the progestins. These hormones essentially hover about waiting for pregnancy, when they perform all kinds of managerial functions. Their name comes from the notion of being involved with pregnancy, as in "progestation." The best known of the group is progesterone. During pregnancy, the progestins help induce fluid retention. That's because the body needs to increase blood volume, providing blood and oxygen for two. They promote weight gain and storage of nutrition. They help stimulate the breasts for milk production. They also protect against too much blood loss when pregnancy does not occur, as the unfertilized egg is washed away in the menstrual cycle. They accomplish that protection by inducing the growth of muscle cells around capillaries in the uterus. That provides additional strength to clamp down on blood flow, preventing a dangerous hemorrhage.

While acknowledging the importance of progestins, I'm going to narrow my focus to estrogens in this case, partly for purposes of comparison with testosterone. The only "female" steroid hormone that even approaches being as well studied as testosterone is estradiol. As with testosterone, the study of estradiol reminds us that no behavior can be explained by a single facet of biology, and when we settle for such an oversimplified explanation—as did the knife-happy gynecologists of the nineteenth century—we get it wrong. The tricky part these days is avoiding total backlash against such judgmental errors. There's a contemporary, very understandable resistance to acknowledging biology's influence on how we act and who we are, much of it in reaction to the abuses of the past.

Consider today's idea that women are more durable, perhaps even designed to outlive men. That's a twentieth century concept. Before this century, so many women died young in childbirth that the cultural assumption was that men were made to outlast women. It even turns out that men accidentally made that gap worse. During the eighteenth and nineteenth centuries, male physicians took over the job of delivering babies from midwives. And since the notions of microbes and proper hygiene were not known, at least until the late nineteenth century, the doctors would hurry straight to the delivery bed from treating patients with raging fevers, often bringing the infections with them. That hadn't occurred with midwives, who only helped deliver babies.

It wasn't until sterile operating rooms, safer anesthesia, vaccines, and eventually antibiotics came into widespread use that the mortality balance corrected—a process that accelerated in the 1920s. Life span improved for both sexes, but especially for women. In industrial nations, women typically now outlive men by about seven years. In the U.S., in 1920, the average life span was 54.6 years of age for women, 53.6 years for men. Now it's 79.9 years for women and 72.8 years for men. Once modern medicine improved the birth survival odds, women began to look almost startlingly stronger. In recent years, women have whittled a little of their edge away by adopting habits that once mainly killed men—cigarette smoking, for instance. But although lung cancer rates are rising faster now in women than in men, the basic mortality difference has remained pretty much the same.

It looks as if nineteenth century scientists had it backwards; they picked the wrong steroid hormone as a promoter of long life. It turns out that testosterone is not that easy a hormone to live with. Testosterone may improve muscle profile and sex drive, it may promote an energetic and competitive attitude, and it may keep men quick in their responses, but all of that hustle seems to be rather exhausting in the long run.

In fact, testosterone can actively depress the immune system. Seattle scientist John Wingfield points out that sperm are alien products in the testes; they definitely aren't designed to stay in the body. Under routine circumstances, a man's immune system might massacre sperm on sight. It appears, though, that testosterone in the testes helps dampen the immune response to the sperm, preventing in-house slaughter. Wingfield says this may have been one of the earliest functions of testosterone. But the problem seems to be that, as with many hormones, testosterone has been adapted for other uses and now circulates throughout the body, trailing its anti-immune qualities with it. Men are notoriously more vulnerable to infectious disease than women. The same is true in other species. If you take the testes out of mice and guinea pigs, knocking out testosterone, their immune systems promptly become more active.

On the other hand, estradiol charges up the immune system. In particular, it appears to stimulate production of two of the immune system's more potent components—killer T-cells, which act to destroy invaders, and B-cells, which are part of the antibody system. It can also, apparently, selectively signal the immune response according to where it is needed. At Dartmouth College, Charles Wira has found that as female rats' estradiol rises just before ovulation, it boots up the immune response within the uterine tissues. That reaction effectively sterilizes the uterus in preparation for vulnerable embryos. But there's no comparable signal to activate an immune response in the cervix or the vagina, and with good reason. There, such a reaction could destroy incoming sperm.

So it appears that one reason estradiol disappears during pregnancy is to allow the immune system to relax, to make the mother's body less likely to attack the embryo. Women's immune systems are measurably less responsive during pregnancy, recovering slowly with the gradual return of estradiol after the birth. That explains why physicians who trailed infection with them in the days before sterile medical practices were especially dangerous to women in pregnancy or during labor.

But estradiol's effects go far beyond the immune system. It summons, demands, builds, and interacts with all sorts of body functions, seemingly affecting everything from how strongly our hearts beat to how quickly our thoughts move.

Estradiol also appears to be influential in such undeniable aspects of good health as having a strong heart. Scientists try to gauge this by, again, comparing postmenopausal women who receive estradiol supplements with those who do not. A study by researchers at Harvard Medical School and Brigham and Women's Hospital in Boston found that estradiol therapy could cut the risk of heart disease almost in half. A recent study by the U.S. Department of Agriculture found that estradiol seems to make blood platelets slightly less active, meaning less prone to clumping and clotting exactly where you don't want them to. There's also evidence, based again on estradiol therapy after menopause, that the hormone increases the so-called good cholesterol levels in the blood. So far, how and why remain poorly explained.

There's a message here in the way menopause keeps creeping into the discussion. That is, that estradiol may be protecting women's health because women are so valuable in terms of species reproduction and survival. In this evolutionary concept, males don't get extensive protection; their critical function is to supply sperm and then they're basically disposable. But not so females, especially in the family of mammals. They have to produce the larger and more complex egg, incubate it, deliver the offspring, nurse the child, and often raise it. They absolutely have to be tough and durable. They have to stick around. Estradiol hovers about like a good-health fairy until a woman stops ovulating, as in (aha!) menopause. And then, suddenly, the woman is not quite so valuable to the continuation of the species. At this point, the theory predicts that the fairy abandons her, that the extra health protections start disappearing. And that's what happens. At menopause, for instance, the incidence of heart disease changes from a mostly male threat to one that targets both sexes equally.

Dr. Mary Haan, who directs a cardiac research program at the University of California, Davis, for the National Women's Health Initiative, summarizes the argument like this: "Women are the main reproductive unit in the species. Many theorists believe that the organism lives as long as it needs to to perform the functions necessary for reproduction. We don't need a lot of men from the reproductive standpoint. That's why they die earlier. And when women stop reproducing, the protections stop for them too." Haan admits to thoroughly disliking the term "reproductive unit" to describe women, and I don't like it either. But, she adds, it works in an objective sense. There's no good alternative theory to explain why female health protections drop away so rapidly after menopause.

There are countless studies on the topic, but I'll cite only one more. Researchers with the California-based Kaiser-Permanente Medical Care Program sifted through the medical records and death certificates of 454 postmenopausal women born between 1900 and 1915. This is a numbers-heavy study, but bear with me; they are interesting numbers. Of those women, 232 started estradiol supplements within three years of menopause and continued the therapy for at least five years. The rest of the women (totaling 222) either refused estradiol treatment or discontinued it within one year. The investigators found that the overall death rate was 46 percent lower for those on hormonal replacement therapy. That effect seemed to get more dramatic with age: of the women in their 60s, the death rate was 16 percent lower for estradiol users; for those in their 80s, it was 54 percent lower.

This does not constitute an unqualified recommendation for postmenopausal use of estrogens. As studying testosterone should remind us, we shouldn't be too quick to slap a halo on a hormone. Neither do we know exactly how to mimic nature perfectly, nor completely understand how the body itself uses the estrogens to best possible effect. There's some visually undeniable evidence of this. For instance, there are some chromosomal disorders in which very little estradiol is produced. An example is Turner's syndrome, in which a girl has only one X chromosome. The girls look like girls—although often unusually small—but at puberty, they don't have enough of the hormone to push them into adolescent development. Physicians usually treat this by adding in synthetic estrogens, delivered by a patch, to induce a stronger hormonal signal.

However, when they first tried this treatment, doctors boosted the girls right up to a normal adolescent estradiol level. It didn't occur to them that they needed to mimic the natural arc of the hormone's increase out of childhood; that the body has precise plans for how much estradiol is needed when and where. That became rapidly obvious, though. The unfortunate girls developed breasts all right, but they were shaped like tubes. The current treatment tries to elevate estradiol gradually, in a rising curve as close as possible to the natural one.

There are three key points in that short story: Estradiol acts as a growth promoter; it's tightly tied to a set schedule of reproduction; and playing with that schedule—in young girls or older women—poses risks that we are just beginning to realize.

My least favorite example—only because I take it so personally—is delayed childbirth. I'm fairly representative of women who put off having children in order to pursue a career first. There are many other very valid reasons why people have children late in life, from a relaxed approach to finding a partner, to a long, difficult path to pregnancy. My choice had to do with my early love affair with science writing. I wanted the freedom to dash about the country covering shuttle launches, volcanic eruptions, visiting lonely radio telescopes at night and flying over gleaming Alaskan glaciers during the day. Can you do that while raising children? Of course you can; you just leave them behind. But if you're like me, once you have kids, you start thinking maybe you can do the telescope story on the phone; maybe you don't need to see the volcano. Maybe you'd just rather be at home. I knew I'd be like that. I waited until I was 34 before getting pregnant with my first son; I was 38 when I became pregnant again. I happen to think that it's a terrific—okay, exhausting too—but still a terrific time to be a parent.

I was truly resentful when, after the fact, a whole series of new studies showed that delayed childbirth increases the risk of breast cancer. The American Cancer Society reported in 1995 that women who have their first child after the age of 30 face twice the breast cancer risk as those who give birth before age 20. Another analysis named it among the top three risks for breast cancer. Did I want to know this? How much unpleasant reality do I actually need in my life? And yet, the undeniable fact is that estrogens promote rapid cell division in the breasts—we know that from the way our bodies change shape in puberty. "If we're not going to accept that estrogen relates to breast cancer, then we need to forget everything we know about the disease," says Graham Colditz, of Harvard University.

It appears—and physicians are still sorting this out—that there are times when such a growth push is more dangerous than others. There are times when it is likely to cause a scramble of cell division with a high rate of error. In pregnancy, a rapid shift in estrogens occurs—the drop in estradiol, the sudden appearance of estriol. You might expect that estriol—so closely associated with a time of rapid body change—would be an especially potent promoter of cell growth. The question, the one I dislike but suspect may be legitimate, is whether women's bodies are adapted to handle that potent shift better at a younger age, perhaps based on an evolutionary history of early childbirth.

It follows, out of that same concept, that supplementing with estradiol later in life might also raise the cancer risk. Postmenopausal estrogen-only treatments are directly linked to increased odds of breast cancer. Harvard University scientists reported in 1995 that for women in their 60s, the risk of breast cancer doubled among those using estrogen replacement therapy. The same treatment seems to raise the odds of uterine cancer as well. Again, the suspicion is that we aren't yet clear on how the body handles the hormones, so that adding a premenopausal estrogen into a postmenopausal woman causes some problems.

This seems particularly true in reproductive tissues; by contrast, there's evidence that estradiol supplements offer some protection against colon cancer. Is there an evolutionary shadow here? Probably. It seems likely that our species evolved with women bearing children at a very young age. It's not just the breast cancer studies that tell us that; studies of the eggs themselves, which seem to degenerate with age, raising the risk of birth defects, tell the same story. In the distant past, many women didn't live through menopause; there's nothing in our history to say that we should be perfectly prepared for a strong dose of estradiol at 80.

On the other hand, there's no particular "rightness" to the habits of our ancestors. Suspecting that we're breaking the patterns of the past doesn't mean we're committing a sin. Are we going to argue that somehow early hominids were morally correct to have children young and keel over early? The point is that modern medicine may identify the risk; evolutionary biology may help explain why it's there. And both disciplines together say—at least, to me—that we should put more resources into this area of research.

If women, and presumably their mates, want to wait for children, we need to work on making that safer and easier. Not all the answers are difficult either: breast-feeding—which suppresses estrogens—appears to reduce the cancer risk by as much as 20 percent. And there's good evidence that adding progesterone to the hormonal therapy mix reduces breast cancer risk. There are some new drugs, still in test-tube stage, which can apparently direct estradiol to a specific target—say strengthening bones in elderly women—while blocking the more dangerous side effects.

We need, beyond that, to keep breast cancer in perspective. Heart disease kills 480,000 women in this country a year; breast cancer kills far fewer, under 50,000 annually. That doesn't diminish the tragedy of the disease or the need to take it seriously. Hormonal replacement therapy and childbirth, early or late, are individual decisions. A woman who has breast cancer, survives it, and then goes into menopause needs to weigh estradiol therapy very differently from one with no such history of the illness. All statistics tell us is that for most of us, the benefits of estradiol outweigh the risks.

Scientists are still sorting through the ups and downs of estradiol's effects on brain function. There's consistent evidence that estradiol levels can be a factor in Alzheimer's disease, for instance.

Here estrogens seem to be dubious guardian angels. Of the four million Americans affected by Alzheimer's, experts say that some three-fourths are women.

It may be that such vulnerability is related—at least partly—to the way estradiol rises and falls during a woman's lifetime. Men get a low and fairly steady dose of estradiol. But women are jacked up and down by the menstrual cycle, buffeted by the changes of pregnancy, and finally tumbled into a menopausal free fall. It appears that the last drop is the most damaging, that some protective effect is lost at that point.

Researchers studying older women have proposed that connection. One large survey, for example, looked at 8,877 residents of a retirement community in Southern California. Two researchers from the University of Southern California, Annlia Paganini-Hill and Victor W. Henderson, examined the nature of the deaths of those residents, concentrating specifically on those who had suffered from senile dementias and those who had not. They found that among the women who died, those receiving estrogen therapy had a 30 percent lower risk of developing Alzheimer's. A 1996 study found that even one year of estradiol supplements after menopause lowered the risk of Alzheimer's.

In fact, Stanley Birge, a geriatrician at the Washington University School of Medicine in St. Louis, has described hormone-replacement therapy for women as "the most promising thing that's happened so far in Alzheimer's disease." Estrogens, he says, have the potential to prevent the great majority of Alzheimer's cases if the early findings are supported by ongoing clinical trials.

Why is that? What is it about estrogens that might buffer the brain so well and, by their decline, leave it so vulnerable? Animal research indicates that estradiol improves blood circulation in the brain, keeping up the health of neurons. The hormones also appear to stimulate nerve cell growth and branching in the exact regions of the brain so damaged by Alzheimer's. For example, it appears that estradiol fosters the growth of neurons within the hippocampus, the region of the brain so closely associated with learning, memory, and spatial reasoning. Brain-imaging studies show that most Alzheimer's patients suffer visible damage in the hippocampus—an erosion of the synapses that allow one neuron to talk to another.

Animal research also supports this theory. If scientists take the ovaries out of a female rat—essentially cutting off estradiol production—neurons almost immediately start to vanish from her hippocampus. That can be reversed just as quickly by giving the rat an estradiol injection. And rats with normal levels of estradiol seem smarter. They do better on tests—such as avoiding electric shocks—than those in which hormone production has been shut down.

Another study, by Catherine Woolley of the University of Washington, found that the hippocampus of female rats seemed very responsive to the normal estradiol cycle. By taking a series of brain images, Woolley demonstrated that dendritic spines—fine little branches that reach from synapse to synapse—grow and retract continually during the cycle.

Female rats have about a five-day estrus cycle. During that time, the number of functional spines increased by as much as 30 percent, falling away as estradiol dropped. Woolley emphasizes that those numbers do not mean that during estrus the rats are suddenly more brilliant animals. She does offer, though, a possible evolutionary background to the changes: the peak dendritic branching corresponds with the peak in fertility. Female rats do more active mate-hunting during this time; they may travel farther from their burrows, stay out longer, and depend more on navigational skills.

The estrogen roller-coaster ride raises some interesting questions. If the levels are constantly changing in mammalian brains, and if, as the studies suggest, this directly affects neurons in the brain, then do we compare to rats on this point? Does the estradiol fluctuation produce a subtle ebb-and-flow in abilities? Do women possess a different brain at midcycle, say, than at other times of the month? Does that explain why we may feel smarter on some days, less capable on others? Do we have any control over that? Estradiol is not comparable to testosterone in responsiveness; it doesn't rocket away every time a woman sees a sexy man, nor does it rev for competition, at least as far as we know. Its influence is clocked by the egg. The timing is not flexible. Ovulation is exquisitely managed, every nuance carefully controlled by one hormone or another.

The ovulation process follows a 28-day cycle, although, as we all know, this isn't entirely reliable. Still, for purposes of explanation, I'm going to stick with the basic pattern.

The 28-day cycle is divided evenly into two phases, called the follicular phase and the luteal phase. The two-week follicular phase begins when the hypothalamus sends an order to the pituitary gland, via a messenger called gonadotropin-releasing hormone. When it receives that message, the pituitary makes another signal, producing follicle-stimulating hormone (FSH). The follicles are tiny, tubelike structures in the ovaries. They contain undeveloped eggs. FSH both nudges an egg into rapid maturation and notifies the ovaries to increase production of estradiol. The climbing level of estradiol during these first two weeks relays a signal back to the pituitary to make another hormone, called luteinizing hormone.

That hormone stimulates actual ovulation—the release of the egg through the fallopian tube and down into the uterus. This second two weeks of the cycle is known as the luteal phase. In the first part of the luteal phase, the body is preparing for the possibility of pregnancy; the lining of the uterus begins to thicken, readying a home in case the embryo implants, and the blood vessel supply to the uterus increases, preparing a food pipeline. After release of the egg, the luteinizing hormone surge causes the follicle to reorganize itself dramatically. It becomes a structure known as the corpus luteum, which begins pumping out progesterone. The luteal phase is named after this structure.

The rise in progesterone triggers a second estradiol surge, peaking about midway through the luteal phase. If the egg is not fertilized by this time, then progesterone levels start to fall. That drop sends a signal to the ovary to cut back on estradiol production. It's at this point that bleeding begins; the hormonal support for the thickened lining falls away and so does the lining itself. It's mostly the cells of the lining, along with some thick, mucuslike secretions, that we lose in the menstrual period; total blood loss averages between four and five tablespoons. (In addition to progesterone helping to staunch blood flow at this point, the estradiol drop is also well-timed. There's some preliminary evidence that estradiol encourages blood vessels to dilate—good for overall circulation, but less desirable during a menstrual blood flow.)

This beautifully orchestrated cycle begins, of course, in puberty. And to refer back to testosterone for a minute, we know that it's in puberty that testosterone and estradiol share some common goals. As they rise in adolescence, they push the child—male or female—toward reproductive readiness. The changes go beyond the obvious physiology of new curves and menstrual cycles. Behaviors and attitudes change too, in well-known ways. Boys no longer see girls as quite, let's say, the pests of their grade-school years. And vice versa. We may not understand all the behavioral influences of the steroid hormones, but there's no doubt that when their tide rises in puberty, desire floats up with it.

There's some very new, very compelling evidence of the link between estradiol and desire in monkeys, from a study by Kim Wallen and a fellow psychologist at Emory, Pamela Tannenbaum. Among the rhesus macaques at Emory's Yerkes primate center was a group of ten adult females, all ovariectomized in other experiments. As a result, these monkeys had hardly any circulating estradiol. The females were a tight-knit group; they'd lived together for six years.

The scientists decided to reintroduce them to a male-female balance. They chose to house the females with ten adolescent males who were about 3.5 years old (in rhesus terms, just approaching sexual maturity). They expected that as the male monkeys hit their sexual stride, the group would be comfortable with one another, forming into the usual interactive society. At that point, they planned to encourage mating by giving the females estradiol, mimicking the monkeys' normal cycle. Meanwhile, they waited for the animals to become friendly. And waited. And waited.

"Contrary to our expectations and, we suspect, anyone else's who would have done this manipulation, the males did not integrate with the females," the scientists wrote. Surprised and curious, the researchers decided they could afford to wait and wonder—how long would it take the monkeys to sort themselves into a social group? After all, rhesus macaques are intensively social. Two and a half years later, the male and female monkeys were, essentially, still not on speaking terms. "The only social interactions between the males and females were aggression by the females and submissive behavior by the males." The monkeys had settled into a state of armed neutrality: two separate groups, with all of the females outranking all of the males.

"Thus, continuous daily contact with females was not sufficient to socially integrate the males into the female's group. Had this been a free-ranging population, there's little doubt that these males and females would not have associated with each other at all," Wallen says. What the researchers found, though, was that estradiol changed that. In weary frustration, Wallen and Tannenbaum decided to inject the females with estradiol for a seven-week period. It only took three days, however, for a standard society to begin organizing itself.

The injections began on a Friday, and by the following Monday, females were grooming males. Within a week, the animals were mating. The researchers watched the group for another two years, taking the females on and off estradiol. In the process, they discovered another aspect of the estradiol injections: estradiol clearly was essential to female desire. The females were only interested in copulation when they received the hormonal supplement. But the hormone was also surprisingly critical in establishing male-female relationships. And after the first estradiol treatment, the females' attitude toward the males changed permanently. With or without the estrogen, afterwards the females still groomed the males, sat by them, and (with the exception of a few males who never became comfortable with the females) built a regular rhesus society. In fact, whereas previously all the males had been submissive, a classic male hierarchy quickly established itself, with two of the males outranking all females.

"These studies illustrate the tremendous power that the female ovarian cycle has over the character of rhesus society," Wallen and Tannenbaum wrote in summary. Does that translate into human behavior? Would women ignore men if they weren't coaxed into interest by estradiol? We probably aren't nearly as hardwired by hormones as that, but in a less overwhelming sense, can such hormones affect our relationships—with men, with other women? There is reasonable evidence from human studies to suggest estradiol does influence attitudes and capabilities, and, perhaps indirectly, the way we relate to people around us.

Rosemarie Krug, a psychophysiologist at the University of Bamberg in Germany, recently investigated the way women deal with feelings of jealousy and frustration during the menstrual cycle. Krug expected to find that women, in a high-estradiol mode, would be more sexually jealous than usual. She asked them to listen to tape-recorded stories about unfaithful men. She expected estradiol to correlate with an angry reaction. What she found, though, was more subtle than that. The women were most angered by the results of the cheating—that the fictional man wasn't spending as much time at home, that he wasn't there to help and to care. In other words, one of the worst things a man could do—at least to woman in estradiol peak—was act like a really lousy potential father. Krug found as well that women were more assertive when estradiol was running high. If she assigned the subjects some problem-solving tests, they were unusually quick to come up with solutions. They were also reluctant to depend on others for help. In fact, she said, they seemed more sure of their own ability to fix things.

Karen Berman, of the Clinical Brain Disorders Branch of the National Institute of Mental Health, reported a similar connection between estradiol and capability. Berman's study involved experimentally keeping both estradiol and progesterone out of the brain. She gave women the drug Lupron, which shuts down production of both hormones. When she scanned their brains using PET technology, she found that activity in the frontal lobes had slowed to a near stop. If she kept the women on Lupron but also gave them synthetic estradiol and progesterone, the brain immediately tuned back up. She measured rapid metabolism as the women did simple tasks, such as card sorting. Berman suggested that perhaps when women are flush with estradiol, they are just a little smarter: "It suggests that during a woman's cycle, she may have opportune times to learn or may be more effective at certain tasks," Berman concluded.

One of the strongest connections between estradiol and intelligence comes from Uriel Halbreich, a professor of psychiatry at the State University of New York at Buffalo. Halbreich compared postmenopausal women of an average age of 52 with women in their mid-30s. He tested them on reaction time, mental alertness, coordination, dexterity, and verbal ability. The younger women outperformed the older. Then Halbreich gave the older women an estradiol boost and suddenly they were clicking in time with the mid-30s group.

Halbreich has also found that estradiol offers some protection against depression. He suspects the hormone may help regulate the neurotransmitter serotonin. Of the postmenopausal women he studied, those who suffered most from depression had unusually low serotonin levels. If, instead of giving the women Prozac (which relieves depression by increasing serotonin uptake in the brain), he treated them with estradiol, he got the same effect—their depression began to fall away.

There's some complementary animal research on this connection, which hints that estradiol's effect on the brain is not so different from Prozac's—that it encourages the brain to be more serotonin-responsive. In an experiment with ovariectomized rats, George Fink, of the Medical Research Council's Brain Metabolism Unit in Edinburgh, Scotland, found that if he treated some of those rats with estradiol, they sprouted new serotonin receptors in the brains within a day. He also found that when the hormone was injected, he could measure increased expression of the genes coding for serotonin receptors. The activity was noticeable in parts of the brain associated with emotional response. Fink has begun to argue that estradiol may be "psychoprotectant," a natural buffer against mental illnesses such as depression.

If so, if estradiol is all about being happy, why do we even have the question of premenstrual syndrome, the notorious mood swing? Where did the idea even come from?

There are some good numbers to suggest, at least, that estradiol jumps and plummets far more dramatically than testosterone. Of course, testosterone has a daily cycle. It's lowest at midday and highest in the night; the fluctuation averages between 100 and 150 percent. In women, testosterone also rises with the ovulation cycle, and the increase is comparable to the daily cycle in men. But, comparatively, estradiol fluctuates in much wider arcs. In some women the increase from base level to estradiol peak has been measured at 4,900 percent. That's the highest variance recorded. But even the smallest increase runs about 650 percent. "This is kind of interesting as it suggests that both males and females may experience about the same percentage change in testosterone (with males experiencing it more frequently than females), whereas only females experience the dramatic steroid changes they do in estradiol," Wallen comments.

Let's look at PMS in that whipsawlike context. Some 150 symptoms have been lumped into the PMS catalogue. They seem to cover the entire range of human health problems, including dizziness, anger, hoarseness, nausea, muscle aches, memory loss, and acne. It's the irritability factor, more than depression, that seems most prominent. In a curious way, PMS—for whatever reason—makes women a lot more like men—more aggressive in their reactions, more likely to snarl, more edgy. So far, the American Psychiatric Association has approached that collection of symptoms tentatively. The association gives premenstrual syndrome the rather stuffy name of late-luteal phase dysphoric disorder, and places it in the appendix of its famed diagnostic catalogue, the Diagnostic and Statistical Manual of Mental Disorders (DSM), rather than giving it full-fledged status. The association does classify PMS, however, as an insurable disorder, meaning that treatment may be covered by medical insurance.

PMS really took off as an issue in Britain (where it is also called premenstrual tension) in the 1970s. At that time, physician Katherina Dalton wrote a best-selling book called *Once a Month*, in which she argued that once a month women, well, lose it to hormones and become potentially homicidal nuts. Dalton became a persuasive expert witness in a number of murder trials. In one, a woman smashed her former lover into a telephone pole with her car. She was tried for murder, but, thanks to the PMS defense, received a conditional discharge from prison and a permanent driving ban. Another woman, a barmaid who killed a coworker in a fit of jealousy, was put on probation and ordered to undergo progesterone therapy. Dalton cites appalled husbands, complaining that their nice little wives, normally so good about dressing pretty and cooking dinner on time, become recalcitrant and obnoxious once a month. It's enough, reading these complaints, to make one wish that the women would suffer permanent and real PMS, because apparently, without it, they're the Stepford wives.

Dalton, who has campaigned for PMS awareness since the 1950s, suspects that falling progesterone causes this personality transformation. She recommends boosting that hormone. However, that recommendation has not held up well. No progesterone mechanism has been identified, and two clinical tests of progesterone supplements proved only that they didn't seem to work. No one, in fact, has been able to explain PMS well. Others have suggested falling estradiol is the culprit, since that seems to be connected to a drop in serotonin. Rising estradiol has been suggested too, since some studies show that it can bring down beta-endorphins, the body's natural system to induce relaxation and pleasure. There are also theories based on prolactin deficiency, vitamin B deficiency, glucose intolerance, and nutritional deficits—in other words, no one knows. That doesn't mean PMS is a myth, however, and I'm among those who aren't about to dismiss it as such.

Provocative circumstantial evidence hints at the estradiol-serotonin link. This has to do with the use of Prozac to treat PMS. In a recent study financed by Eli Lilly, the drug's manufacturer, 405 women who had long reported PMS symptoms volunteered to participate in a trial in which half received Prozac and half a placebo. The trial was a "blind" one—none of the women knew whether they were getting the real thing. Did the results solve the PMS mystery? No. But half the women getting Prozac reported at least a 50 percent reduction in symptoms. Of those getting the placebo, less than a fourth reported that kind of improvement. The fact that women on sugar pills reported any improvement is a reminder of the famed "placebo" effect: believe that you may be receiving a good treatment and the belief may make you well. But clearly, for a fair number of women, Prozac helped. And since the drug boosts serotonin uptake (and estrogen can pull it down when it falls), this could be counted as supportive evidence.

The real problem with PMS, I think, is that it's been exaggerated into a concept of total loss of control, as with the cake-wielding lunatic wife. Yet the solidly consistent symptoms are far less than that—mostly irritability and impatience. And a little irritability is not a mental illness, which everyone, male and female, should appreciate. Of all the women who report some kind of menstruation-related mood swing, only a very small number—about four to six percent—say that it's truly distressing. These are the women who may, indeed, need Prozac once a month. We each have our individual biology; hormones and drugs—and ragweed, for that matter—affect every person in a different way. We might predict a very different hormonal impact for the woman who generates a 650 percent estradiol increase compared to one who spikes a 4,900 percent increase.

Either way, PMS does not equate with lunacy. The popular insistence on portraying it as such does nobody any favors—not those who may suffer from it, not women in general. Feminists, understandably overreacting to the stereotype, argue that the whole PMS notion is a carryover from the days of ovariectomies, from the time when medicine, and culture, could portray women as emotionally out of control—and, therefore, in need of control. Susan Faludi used PMS as an example of men putting down women in her 1991 book *Backlash.* She complained that placing the syndrome into the DSM was tantamount to defining normal female chemistry as the hormones of craziness.

British science writer Gail Vines, who tends to mock what she sees as an obsession with hormones and behavior, points out that many of the PMS studies have looked at homemakers—women stressed out by the constant demands of small children. Perhaps, Vines suggests, they feel trapped; perhaps their lifestyle, with its "inescapable commitments," may put them already on an edge of irritability and fatigue. American sociologist Anne Campbell makes a similar point—that the hormonal flux simply strips bare the fact that many women are unhappy in a world in which they often feel both powerless and dismissed.

I don't entirely accept that. It seems possible that estradiol does jolt the system; it certainly rockets to high levels at extreme speed. We've evolved with that monthly liftoff and, undoubtedly, adapted to it, meaning that most of us just get irritable. And some of us—those with a more sensitive biology or a sharper fluctuation—may suffer stronger effects. There may be a potent biological reason behind the misery claimed by that small number who say PMS wrecks their lives. It's easy enough to accept that if you don't see PMS as a political symbol.

Accepting a background biology doesn't make the arguments of Vines and Campbell ridiculous, either. If a woman is prone to being knocked off-balance by a hormone shift, and if that woman is also unhappy, then her dissatisfaction could certainly be boosted by the change. In that sense, there may be something to the idea that PMS, with all its associations of red rage and revenge, also speaks to a more generalized anger or frustration among women.

Scientists, in fact, have begun exploring just that concept in depression research.

Almost twice as many women as men seek professional help in industrialized countries (the exact U.S. statistic is 1.7 women for every man, but the basic 2:1 ratio persists across many countries, including Canada, the former West Germany, Korea, and New Zealand, obviously a wide range of cultures). For many years, that difference was explained away by the idea that women were more hormonally unstable. Recently, the search for an explanation has begun to broaden, to look beyond reproductive biology. This doesn't dismiss hormonal factors—the female depression gap shows up only after puberty, timed too coincidentally to the rise of estradiol. But the latest analyses of depression take a wider-lens approach, acknowledging biology as an interconnected part of broader destabilizing influences that may also include troubling aspects of relationships, status, and power.

Susan Nolen-Hoeksema, at Stanford University, has spent the past decade focusing on the differences in depression between adolescent boys and girls. She provides some reasons why we shouldn't assume that the hormone story is simple. One is that, although estrogens and progesterone actually begin rising before visible puberty, at about age nine to ten, the visible gender difference in depression appears at least two years later, thus discouraging the notion that it's a straight hormonal side effect.

Second, in puberty, estradiol can change a girl's world without any direct influence on behavior. What happens when your whole body shape changes and suddenly you're no longer a stick figure drawing in a child's landscape? Other people see you differently, they change the way they treat you, and you see yourself differently. A lot, including how you treat yourself, may depend on whether you like what you see.

From about age nine to age 13, girls gain an average of 10 inches in height and 24 pounds of body fat. By comparison, boys are on a slightly delayed track. They begin puberty about two years later than girls. And their main growth is in muscle and bone, not fat.

Many girls hate these results, as Nolen-Hoeksema points out. They stare at themselves in the mirror, seeing unwanted padding, while boys are suddenly flush with body and power. A host of studies shows that teenage girls are inclined to dislike the way they look. Boys are usually satisfied with themselves—at least according to surveys.

Is there something about our society that creates this particular gender difference—male satisfaction, female dissatisfaction? We discussed earlier some of the biological theories as to why female appearance may be so important to males. We also discussed reasons why male appearance might matter to females. So why, in female perceptions, is the body you have never good enough? There are long-standing feminist arguments that our culture's fashion ideal is really about male power: setting a goal of nearly unattainably polished looks, thus making sure that most women feel unpolished and inadequate. It is interesting—or maybe just frustrating—that in a culture which makes being thin part of being good-looking, biology makes it easier for men. Testosterone doesn't pack on the fat the way estradiol does. If anything, today's fashion extreme asks women to have bodies more like men.

California anthropologists, such as Robin McFarland of De Anza University in Santa Cruz, have recently demonstrated that in a powerful way. McFarland reported at the 1996 meeting of the American Anthropological Association that female bodies are, in fact, beautifully and functionally designed to store more fat than male ones. This is not only true for humans but for other primates, from African gorillas to Indonesian pig-tailed macaques. Fat storage seems particularly tied to lactation, and researchers have found that when women nurse, it's the fat from those notorious "problem" areas—hips and thighs—that is accessed for the milk supply. It's estradiol, of course, that stimulates the accumulation of such fat supplies, and in turn, it seems that an adequate fat supply may help maintain a healthy level of the hormone.

Scientists have found that very thin women tend to have low levels of estrogens and, also, difficulty in maintaining bone mass.

As mentioned, estradiol also helps regulate bone mineralization. And, according to University of California, Santa Cruz, anthropologist Alison Galloway, this again may be tied to lactation: the hormone may induce the bones to release calcium for breast milk at the appropriate time. "I suspect that estrogen and fat and bone form an important triad," she says. For instance, Galloway notes, women of European descent find it much harder to lose fat than those with an ancestry from warmer climates. They also are more vulnerable to bone demineralization. She suggests that this might result from an environmental adaptation—that when early hominids migrated from Africa into icy northern climates where food was far less abundant, the only source of nutrition for children, at times, might have been breast milk. So women of northern European heritage might have an evolutionary history that predisposes them to lose bone minerals more easily (making them more accessible for lactation) and to store greater amounts of fat.

"If you look at the bodies of fashion models, except for the breasts, they are really male bodies," McFarland points out. "Long, stringy, ribs showing, narrow hips. It's silly for women to want to achieve that. We need to come to grips with the understanding that women accumulate fat easily and that's not necessarily bad."

On the other hand, once you exit the parallel universes of *Vogue* and Hollywood, most people don't actually demand such extremes. In a straightforward experiment to test such preferences, John Krentz, of Hanover University, set up a "female body" Web page, filled with outlines of women's figures. He asked some 500 Internet visitors, both male and female, to choose their favorite shape. Both sexes chose a moderate, curvy shape over the somewhat androgynous fashion-model look: "Personally, I don't find models interesting," Krentz said. "They're sticks."

Someone else's reassurances, though, don't help that much if you're a teenage girl longing after willowy androgyny. Depression can begin with just this mirrored misery, Nolen-Hoeksema points out. She suspects that body image may be one major issue that puts boys and girls on differing paths toward self-respect. But adolescence changes more than outward appearance. Before the testosterone surge, boys tend to be more negative than girls about life. In one study of third-grade children whose parents had divorced, the boys were more depressed by the split; the girls were more convinced that life would get better, that relationships would repair themselves.

In the early teens, though, girls show more pessimism. They become more withdrawn, mulling over their problems, internalizing them. This isn't always a good way to handle trouble; Nolen-Hoeksema's work suggests that silently turning over a problem, not sharing it, makes people even more pessimistic. In adolescence, boys make their stresses public. They act out; they're often disruptive in class and confrontational elsewhere. As a result, boys get more help. "Boys with conduct disorder are a real pain and they get a lot of attention," Nolen-Hoeksema says. "Depressed girls are quiet and unassuming and they don't bother anyone."

She points out that parents often pressure girls into that role, especially as they move into the teen years. Before adolescence, parents tend to be much more tolerant of tomboy behavior in girls. As I said, it's a lot easier to be a sports-loving girl than a doll-loving boy. But as children grow older, parents tend to set stricter rules for girls. They're intolerant of teenage girls who want to behave like rowdy, adolescent boys. It seems to be a time when parents expect their sons to act out, but they want their girls to start acting more like "ladies." They're increasingly protective of girls as well. Teenage boys are allowed more independence, more permission to go out alone. Unfortunately, as a statement about society, there's some basic common sense to these attitudes, at least concerning safety. A young girl out alone is vulnerable in ways that boys are usually not.

But research suggests that these parental attitudes are not merely protective instincts at work. The parents questioned said they wanted to encourage their sons to explore; they actively tried to help build their confidence because their expectations were higher. They predicted greater career chances. They tended to underestimate their daughter's abilities. A 1990 study found that if parents were asked to rate their daughter's abilities on math tests (where expectations for boys run high), they actually marked them lower than the girls' teachers did. At the same time, they tended to be overconfident about their sons' capabilities.

Moreover, girls with good academic records tended to be rated as unfeminine by their peers. No surprise there—it's about at this point, in adolescence just after puberty, that an academic gender gap becomes really visible. Boys begin outshining girls in school. Parents and teachers begin to rate boys as smarter. And girls, traditionally, at least, have been caught in a peculiar kind of trap. Parents push them to be more meek, peers to be more dumb. Nolen-Hoeksema and her colleagues, in a survey published in 1994, found that girls who accepted such a narrowing of the world to the stereotypical feminine role tended to be at higher risk of depression.

In another analysis of gender differences in depression, researchers noted that teenage girls are far more easily upset by relationship problems than are boys. This again marks a reversal from the more carefree prepuberty days of females. Remember how needy young boys seemed? Everything changes after puberty; girls want more emotional support. A four-year study of teenagers in Iowa found that of children with cold and unsupportive parents, girls became depressed far more rapidly than boys. The reverse effect was also true. Girls seemed to benefit more from warm, loving families. If a boy was diagnosed in the early stages of a depression, attention from involved parents didn't seem to help him as much as it did a comparable girl.

That difference reinforces an earlier point, that females build—and maybe lean on—stronger emotional connections. That doesn't go away in adolescence. But puberty turns out to put a major stress on those connections for girls. They're suddenly very interested in establishing connections with boys. After a childhood of gentle, chatty relationships with other girls, they're suddenly learning—or trying to learn—to build relationships with boys, who are neither chatty nor gentle, especially in the teenage years. The energy girls put into attracting those boys is energy drawn out of their friendships with other girls. The once-reliable network becomes less so.

It raises an interesting question about whether hormones, theoretically designed to drive relationships with the opposite sex, might change relationships within members of one's own sex. Desire is in some ways a Trojan horse, dividing once-united females and pushing males into unfriendly competitions.

Tannenbaum and Wallen have explored this issue by studying the role of estradiol in rhesus macaques, exploring the relationship between a hormone and, if you will, friendship. Rhesus females take care of each other and each other's children. Overall, they bond with each other far more closely than they bond with males; this is a polygamous society and there's no indication of a male-female partnership in raising the young. So in terms of survival, it's the bond with other females that's most important. And yet, in the ovulation cycle, as estradiol rises in the follicular stage, females start edging away from each other and begin pursuing males. The Emory researchers found that they could informally track the rise of estradiol by observing the number of females sitting closer and closer to males.

A female rhesus macaque does the courting, and sometimes it takes her hours to seduce the male of her choice. The pursuit comes at the expense of time that had been spent maintaining her female friendships. As estradiol rises, female rhesus macaques cut down on the amount of time they spend grooming other females and sitting with them. And the strain shows. During the mating season, females are unusually hard on each other. There's threatening body language, more follow-through aggression, and more submission by lower-ranking females. As the scientists say, "Thus the female experiences two opposing forces, her own sexual interest in the male and the negative interaction with other group females. It's interesting that the hormonal changes that produce increased sexual interest in the male also switch female affiliative [bonding] behavior away from other females and to the male." As soon as estradiol falls, though, the females quit grooming the males so often and begin repairing their relationships. The study reminded me of a thankfully forgotten saying from high school: "She had a better offer." It meant a particular girl wasn't around, hanging out with the girls, because she'd gotten a date with a boy.

But I'm not suggesting that rising estradiol makes teenage girls reorder their friendship priorities. Science isn't even close to making that association yet. On the other hand, perhaps there is a kind of unhappiness intensified by the social tensions that hormones add to our lives, such as the rise of estradiol in puberty. Ellen Frank of the University of Pittsburgh studies depression in women and she puts it like this: "I do believe that women are genetically preprogrammed to be more affiliative [build stronger relationships]. Interpersonal attachment is a bigger deal for women than men, and that's virtually true in all cultures and times. It has an adaptive significance for the survival of the species. If women didn't attach, babies wouldn't survive. If depression is about loss of attachments, and I suggest it is, then the events that precipitate depression tend to be about interpersonal loss. If we have one half of the human race that's more preprogrammed for attachment, then that's the half that's going to be more vulnerable."

Puberty for girls is, in many ways, a biological preparation for childbearing and rearing. Following Frank's line of reasoning, it may be that as estradiol rises, so does intensity of emotion and connection. For instance, it appears that estradiol can help cue oxytocin, the hormone thought to be so influential in bonding behaviors. In fact, studies with rats show that if estradiol is blocked, then oxytocin does not bring on nurturing behaviors—until the animal is also given estradiol injections. In subtle and still mysterious ways, estrogens seem to help prime women for an emotional response.

Ellen Frank isn't the only scientist who suspects that such emotional intensity makes women more vulnerable. In a project designed to explore that idea, federal researchers at the National Institute of Mental Health asked ten men and ten women to recall the saddest images of their lives. The memories were wide in range but heartbreaking in the retelling: a little boy standing at his father's funeral, a woman who thought she was happily married hearing her husband demand a divorce. The participants deliberately recalled those scenes while the researchers did a PET scan of their brains. In both sexes, the front of the limbic system glowed in response. But in women, the active area was eight times larger than in men. The scientists proposed that this difference might help explain why women more often succumb to sadness. In depression, the limbic system becomes unresponsive and lethargic. The researchers speculated that perhaps the intense female response is sometimes too much, that it wears out the neural circuits. The result is a collapse into numbness which may, if contained, allow recovery.

But Frank argues that human biology is too complex to be explained that simply, partly because we can make choices about our behavior. "I have very substantial sympathy for the idea that there's a lot of genetic preprogramming of behavior," she says. "It's most obvious as you go down the phylogenetic scale. As you move up, though, things get more complicated. We, as humans, have retained enormous amounts of preprogramming, but also we have a massive brain, and that enables us, I think, to escape from the preprogramming, because we have conscious will to do something that goes against what our brain is telling us, because it's the right thing to do or because, dammit, we just feel like it right then."

She continues: "I think for a long time, views about this issue were very polarized. The biologists saw this [female depression] as an obvious example of a sex-linked, gonadal hormone-driven issue. On the other hand, psychosocialists, especially out of the women's movement, saw it as based in the enormous disadvantages for women, such as abuse. And they said, pooh, pooh, we don't need biology; it's obviously a result of women's downtrodden state. We've all come to appreciate that neither view alone is correct."

It's not as if only women get depressed or as if men's depression rates are unchangingly low. Okay, the rates are about two women for every one man now, but that gap used to be a lot larger: the first such comparison in the United States, in 1947, showed an eight-to-one ratio. A follow-up study showed that between 1957 and 1976, gender-related differences in mental distress decreased by 38 percent. Women reported no real change in depression, but men were much more unhappy; it was that difference which closed the gap. In the past 20 years, the suicide rate for young men (between ages 25 and 34) has increased 26 percent; the suicide rate for women in the same age group has dropped 33 percent.

Some researchers suggest this reflects a response to loss of power as greater equality emerges. Others suggest that twentieth century culture has pushed men too far into overdrive, that they feel increasingly isolated by work and unconnected to family and friends. Depression in men serves as yet another reminder, if we need one, that one hormone never explains a behavior. We can hardly argue that men are suddenly brought to a state of despair by estradiol; there's no evidence to suggest the hormone has suddenly concentrated in men to produce harmful effects.

"My own view is that loss attachment, breaking bridges, and feelings of interpersonal inadequacy are the real issues," Frank says. Maybe, indeed, women are hurt more by shifting relationships; maybe our brains burn brighter in despair. But there's nothing in the numbers or the studies that says that men are magically buffered from those feelings. They may handle them differently; estradiol may be less significant. But in exploring depression, we still have an unfortunate state of gender equality: loss and grief and our bodies' responses to them are painful challenges that we all face, separately or together.

•

Chapter Eight

COUNTERSTRIKES

Love, Lust, and Rape

•

Scientists love fruit flies, at least as experimental subjects. Fruit flies definitely don't love them back. If the insects had emotions, they'd undoubtedly hate the scientists' guts. The reason researchers spend so much time with fruit flies, raising great clouds of them in laboratories, is that they've learned to play all kinds of fascinating genetic games with the bugs, scrambling their codes like eggs in a bowl. Fruit flies are easy this way. They have only 8,000 genes, wedged onto four fat, highly visible chromosomes. They live only a few weeks, so that you can fast-forward through generations. And they're tiny, like ground pepper grains with wings, so that you can study thousands of them without needing much more than an extra-large box.

Creating mutant fruit flies is practically an initiation rite into genetics. Change the color of their eyes? Done that. Grow eyes on their wings? Done that. Change their sexual orientation so that males devotedly pursue males? Done that too. In fact, researchers say that the gene (known as *fruitless*) which apparently controls sexual orientation in fruit flies seems to be a "master" gene for male sexual behavior, regulating not only mate choice but the style of courtship, right down to whether or not the male sings during the seduction.

Because of this ability to explore and tweak and ramble up and down the genome, fruit flies have taught biologists an incredible amount about what genes control: when they are less than influential and when they matter a very great deal. The biologists working with *fruitless,* for instance, say that if a master behavior gene exists in fruit flies, there may well be such supergenes in other species, even our own. For all that we tend to dismiss such tiny insects as nuisances and pests, in many ways they have served as a startlingly clear window into the inner workings of life.

In the spring of 1996, William Rice, a geneticist at the University of California, Santa Cruz, published a vivid and in the end deadly study of fruit fly sex. Rice was particularly interested in males and females as part of each other's environment. We must constantly adapt to the world around us, but that doesn't mean just the weather, the landscape, and the local nuts and berries. It means each other, too. Males, argues Rice, are a force in female evolution and vice versa. "The sexes evolve substantially in response to each other, not just in response to the physical environment or to predators," he says. Further, in the stripped-down evolutionary sense of genetic survival, the two sexes aren't necessarily on each other's side: males would seek to spread male genes around, females to further their own agenda. So Rice's study was, in a sense, an inquiry into opposing strategies.

What if one sex could evolve without the other? What if there were no counterbalance? That was what Rice looked at in fruit flies. He created a group of tiny winged studs, which he called "supermales." They were engineered toward an exaggerated masculinity: hyperaggressive in mating, shoving other males aside, prone to fathering only other males. The supermales thus re-created themselves abundantly in following generations. And what did that mean to the females? It killed them off.

Fruit fly semen is normally slightly toxic—and hard on the females' health. The supermales, though, had supertoxic sperm. Rice had genetically bypassed the normal influence of female genes, which tends to reduce the toxicity level of the sperm. The supermales thus came as a horrible surprise to the females' systems; they hadn't evolved any response to this threat, any internal antidote. They died—sorry—like flies. Taken to a worst-case conclusion, the unchecked evolution of supermales could be very hard on overall reproduction—what if the sperm became so toxic that the females didn't reproduce at all? A check-countercheck turns out to be an essential element of species survival. In the short term, though, as Rice points out, the fact that "more females died after bearing their offspring was of no consequence to the supermales, whose sole goals in life had already been realized." The supermales were more successful than ordinary males. They by far outperformed them in creating offspring. And if one female's life span was drastically shortened, there was always another potential mate (although, I suppose, too much toxic sperm could change that as well).

The fruit fly story is a dark little tale of sex and death, but the lesson behind it is not so little. It reminds us that in basic sexual drive, males and females don't always function as partners. There's an underlying competition as well. Let one side's evolutionary adaptation go too far in favor of its own sexual strategies and the result can be deadly for the other—and, if no counterstrategy can be found, potentially deadly for the species as well. Now, if we consider that males and females in species beyond the fruit fly also have separate male and female sexual strategies and responses, and that those species include human beings, perhaps we have yet another approach to understanding why women and men behave as they do toward each other.

If men and women "turn on" differently—and there's a plethora of evidence to suggest that—then what explains it? Is it as benign and uncomplicated as two differing yet harmonious hormonal systems, designed to match up sperm and egg? Or is there a strike-counterstrike element to human sexual arousal? If matters sexual really can be seen as a battlefield, is one sex more dangerous than the other?

Notice that in Rice's fruit flies, the males are harmful and the females at risk. He makes poisons; she defends herself. In a big, socially complex species like our own, this would play out differently. While it's true that frequent, unprotected sex with many partners can be dangerous to both men and women (with transmission of viruses or bacteria), it's not a true analogy to the sexual drive toward winning the gene game. In terms of totally self-absorbed sex—sex with complete and sometimes murderous indifference to the female—the human analogy is rape. As we explore the biology of desire and arousal, the most troubling question is whether the gender differences in this realm represent a facet of evolutionary one-upmanship.

The scientist who has most eloquently argued that theory is Randy Thornhill, an evolutionary biologist at the University of New Mexico in Albuquerque. Thornhill's position—admittedly controversial—is that rape is foremost a sexual act, based in the varying reproductive strategies of men and women. He first published the theory with his ex-wife, Nancy Wilmsen Thornhill, and has remained its most outspoken advocate. The idea is based on the familiar egg-and-sperm explanation of gender behavior. Producing a child requires a major commitment from a woman, beginning with a lengthy pregnancy and continuing with the production of a helplessly dependent infant who will die unless she (or a surrogate) feeds it. Producing a child requires a minimal commitment from a man—the famous four-minute sexual act. A man can walk away unencumbered in ways that a woman cannot. And given the above considerations, a man who walks can produce—in terms of sheer numbers—a lot more children than a man who stays to help (and a lot more children than a woman, even if she tries very hard to keep them coming). The high-end statistical comparison, recorded in the *Guinness Book of World Records*, is a high-ranking Mongolian overlord of ancient China who fathered 888 kids, versus a Soviet woman, back in the eighteenth century, whose offspring totaled 69. (She tended to deliver triplets.) Don't tell me that rape didn't feature in the first number. And do tell me why at some point, after, say, 30 children, the Russian housewife couldn't have developed a permanent headache. But the point is the gap: the 800-plus difference in possible number of offspring and the fact that men are designed to be able to pass along genes so incredibly prolifically and that women are not.

"In humans, there's a large sexual asymmetry in the minimal reproductive effort required for the production of offspring," Thornhill emphasizes. "The minimum for a man is a few minutes of time and an energetically cheap ejaculate; the minimum for a woman is nine months of pregnancy and a long period of lactation." Thornhill doesn't argue that men and women today plan their lives around those estimates. But he suggests that early in human history, among our more primitive ancestors, that difference might have influenced how males and females behaved. If the basic goal is reproducing one's genes into another generation, then men and women might be put into conflict. Males would be most successful if they got many women pregnant, essentially choosing quantity over quality, whereas women would be best served by insisting on quality.

The critical aspect of this argument is that women want to choose their mates, leaning toward healthy men with the promise of being good partners and providers. As a result, some males wouldn't make the cut—they'd be left out of the reproductive cycle unless they forced the issue. So women's choosiness runs in direct opposition to a male drive in favor of multiple partners. The main thing our species has in common with other species where persistent and forcible rape occurs—mallard ducks, elephant seals, orangutans, and scorpion flies being the most notable—is that female choice is an important aspect of sexual behavior. Thornhill argues, then, that rape is an ancient reproductive strategy, essentially, that may be still woven into the background of our biology.

"Because of this sexual asymmetry, during human evolution, males who could gain sexual access to multiple females outproduced males who could not," Thornhill insists. "They were thus favored by natural selection." That would argue for a selective conservation of the genes of men like the Mongolian multiplier. Finally—and this has made Thornhill's argument about as controversial as they get—he argues that males are winning the conflict at this point;

they're on the supermale end of the fight. "You get the evolution of females' resistance to mates they don't want," he says. "You get males who are selected to mate regardless of what the female wants. It becomes an arms race on an evolutionary scale."

Feminist scholars have countered that rape is indeed about conflict, but not about sex. They've suggested that it should be classed as either an asexual criminal act or a power play—men asserting themselves over women. In a response to the Thornhills' first paper, published in the journal *Brain and Behavioral Sciences*, feminist Susan Brownmiller pointed out that rapes are committed with sticks, gun barrels, bottles, and other instruments that tell more of torture than of lust. "Ninety percent of violent crimes—rape, robbery, assault, and murder—are committed by men, and largely by young men at the height of their physical strength and athletic prowess, who prey on those they perceive as weaker and more vulnerable . . . ," says Brownmiller. "The forcible rapist (who is also the burglar, the robber, the mugger) retires from the field when his legs give out. Social policy must address itself across the board to the problem of confused young males who see their manhood in physical aggression."

It's sometimes tempting, when thinking about human behavior, to simply declare the whole subject to be complicated to the point of hopelessness. Rape strikes me this way. Is it rape if a nonsexual instrument is used? If the victim isn't even old enough to crawl? When I was a very new reporter, covering the police beat in northern Georgia, I had to write a story about a 21-year-old father who critically injured his one-month-old daughter by shoving a pair of scissors into her vagina. Do we count that as rape, or rage, or pure psychotic insanity? Of course there is a sexual connotation to the act. But who was he angry with, the child or her mother? I'm not trying to equate an attack on a baby with a man forcing a woman at knifepoint to submit to him sexually. My point is only that neither crime fits into a neat little slot. There's pure violence in both; there's power at stake; and there's the element of sex. They don't separate easily. And I think it's a mistake to define rape too simplistically—as a crime of sex only, as a crime of violence only, or as a crime of hate, or contempt, or lunacy, when it can be any combination of those things and more.

If we want to control or eliminate rape—in the most optimistic of worlds—then we have to be honest about everything that might lie behind it. Thornhill makes several arguments on that front: first, that reproductive biology might indeed predict that men would combine forced sex with other violence (such as a beating) if—and this is a big, unknown if—early human ancestors lived in a dominant-submissive relationship in which males were sometimes forcibly dominant; and second, that there's plenty of violence—including men attacking women—that does not include forced sex. There's undoubtedly an element of sexual tension in most spousal abuse. But some men become aroused by hitting a woman and others don't; they just hit. To overstate the obvious, raping a woman is different from beating her up. We actually need to understand the difference between the rapist and the mugger. Finally, and to Thornhill this may be the most important of the three points, saying that a behavior is biologically influenced does not mean that behavior is acceptable. It doesn't give it some kind of naturally moral rightness. Many species fight and kill each other; we don't take that as a naturalistic endorsement of murder. With rape, Thornhill argues, the only reason to reach for its evolutionary origins is to figure out how to eliminate it. Isn't the point of civilization, after all, to create laws that encourage the best of our natural behavior and try to stamp out the worst?

Thornhill teaches an advanced sociobiology class at the University of New Mexico. One of his goals for the course is to bring his students to a point where they can accept that the potential for rape may be part of human biology: "Essentially, all males, in the right circumstance, respond to forced sex," he says. "That's what I tell my students, but we spend a long time getting there, building background, studying Darwinian theory. And then we say, let's understand this in the context of an evolutionary warfare. This knowledge can improve a male's decision making. It's self-awareness—learn the circumstances that might make you vulnerable to these kinds of behaviors, and improve your free will by knowing where your biases are."

All of this builds on the premise that reproductive strategy—and its bedmate, if you will, sexual desire—is different in males and females. That premise may seem incredibly obvious. But there is an abundant feminist literature arguing that women are perfectly able to be as sexually aggressive—and even impersonal about it—as men. In this light, in fact, Brownmiller expresses frustration over the forced-sex imbalance: "Intrinsic differences in size, strength, and reproductive anatomy permit men to rape but deny an equal opportunity to women." Such insistence is partly pure challenge to a culture that many feel has repressed women's natural sexuality and made us all neurotic with the notion that "good girls don't," while encouraging boys to have plenty of creative sexual experiences. Fighting the double standard makes sense—what's wrong with wanting men and women to enjoy sex equally? But that doesn't mean, I think, that women want to become men. When it comes to rape, I wish men would, instead, act more like women. With that, we could come close to eliminating the problem. I've no desire whatsoever to push my sex into matching masculine achievement in rape statistics. And I'm not insulted by the idea that men are more sexually driven than women overall, which seems to carry a definite plus and minus factor. If we want only to one-up each other in terms of quick sexual hits, we might as well evolve into rabbits.

There's an evolutionary argument for why men become aroused more quickly and easily than women. If a successfully reproducing male needs to inseminate a lot of females, he'd better be wired for response—instant response. It shouldn't take much; by theory we should see a rapid response to the very sight of a sexually attractive partner. Every study shows that this describes males. There's some strong evidence, for instance, that testosterone is wired for visual response. T-levels (for testosterone) rise in monogamous male birds as they watch a female. Female birds, however, are rarely jazzed by mere sight of the opposite sex; in monogamous species, they want song and food and courtship. Or consider a polygamous species: remember that when male rhesus monkeys watch a companion copulate, their own T-levels climb some 400 percent. Of course, the rhesus is female-dominated and the males still have to be courted by females. But, let's face it, they are primed. By the end of the intense rhesus mating season, males often lose up to a fifth of their body weight from sheer, exuberant and extensive sex.

In humans, the male system sometimes seems so jittery with sexual readiness that just about anything—high-heeled shoes, a smile, a friendly conversation—will produce a sexual response. "One of the most telling gender differences can be detected very simply," says Kim Wallen. "Just ask men and women how often they think about sex. What you find in a large population is that the average for men is three to five times a day. And the average for women is several times a week to several times a month. Men are just more inundated by sexual thoughts."

Anthropologists have actually analyzed sexual response to smiling by studying casual conversations at bars. Let's stipulate, as they say in court, that the basic ambience of a casual conversation in a bar may well be front-loaded for sex. Still, women insisted that they smiled at men to encourage conversation, considering it a "Let's talk and get to know each other better" signal. Many men, though, didn't read the smile that way at all; they in turn insisted that it was pure come-on, a signal for "Let's hurry up and screw."

Conversation itself excites men more. In another example of notably different response, researchers played audiotapes for male and female university students. The dialogues on the tapes depicted a whole range of conversations between men and women. Some were wildly erotic, some frankly routine. In one of the latter cases, a conversation involved a couple discussing the merits of an anthropology major over pre-med. The erotic tapes aroused everyone—skin warmed, pulse quickened. But some men also showed arousal to exchanges like the career discussion. In fact, their response was stronger, hotter physiologically, than the female response to the erotic stuff. Can you imagine women getting all hot and bothered by hearing a discussion of potential job choices? It seems, though, that just listening to male and female voices jostle against each other is enough for some men; poised on the near-edge of arousal, they tumble over.

The Kinsey Institute once painstakingly showed a group of men and women a series of photos and drawings of nudes, both male and female. Did everyone get off on Botticelli's Venus and *Playboy* centerfolds? Hardly. But there was a significant gender gap. Fifty-four percent of the men were erotically aroused versus 12 percent of the women—in other words, more than four times as many men. The same gap exists, on a much larger scale, in the business of pornography, a $500-million-plus industry in the U.S. which caters almost entirely to men. In the first flush of 1970s feminism, two magazines—*Playgirl* and *Viva*—began publishing male centerfolds. *Viva* dropped the nude photos after surveys showed their readers didn't care for them; the editor herself admitted to finding them slightly disgusting. A *Playgirl* reader survey found that men liked the nude pictures a lot better than women, leading to widespread speculation—denied by *Playgirl*—that the centerfolds appealed mostly to gay men.

There are arguments that women are simply too repressed by society to become so easily aroused. But Wallen points out that society does its best to repress men's fascination with pornography—to little success: "It seems quite likely that the lack of the female-oriented industry simply reflects the small size of the market," he says. Women, in general, don't get so physically excited by anonymous bodies and sex among strangers. Evolutionary biologist Donald Symons, of the University of California, Santa Barbara, nicknames men's response to pornography "pornotopia," a dream of that perfect land where sex is always easily available and unencumbered by emotional baggage. Symons argues that the comparable big-bucks industry for women is the production of romantic books and movies; they capitalize on women's fantasies of sex together with love and commitment.

In an exploration emphasizing just that difference, scientists read sex stories to a group of college students. The plots deliberately combined touchy-feely romance and steamy sex. When the subjects were asked what they recalled in detail, guess who remembered what? Women recalled lines such as "They looked deeply into each other's eyes." Men, on the other hand, accurately recited lines such as: "She clutched his back and wrapped her legs around him." Of course, there was some crossover; not every woman was jazzed by staring, not every man by clutching. But the basic dichotomy plays out. In similar studies of how people view erotic films, psychologists found that women prefer ones that build to a relationship, slowly evolving into sexual relations. Men like those the least; they complain that those stories take too long to get to the—you know—point.

As Thornhill says, the fact that men are so readily aroused by photos and innocent words, and often so unaroused by the trappings of romance, makes perfect sense in the evolutionary picture of men revved for multiple encounters versus women being a lot pickier, looking for help and commitment. But as he also points out, the same disposition that makes career discussions sexy also predicts that women don't have to be responsive, or even remotely interested, for a man to become sexually aroused. And it suggests that many men, at a fundamental level, will find the extreme form of noncommittal sex—rape—sexually arousing.

In fact, there is a series of studies of university men that tend to back up Thornhill's argument. Again, these involve listening to audiotape descriptions of sexual encounters, in this case tales comparing rape with consensual sex. And let's give scientists credit for intelligence at this point. They don't just ask participants to raise their hands and say, "Gee, I really got off on that one." Usually, these studies involve complex measurements. Participants do write down their reactions, but researchers also monitor increasing heart rate, quickened breathing, skin temperature changes. And sometimes they're even more direct. They use equipment to measure swelling and erection. A study done by psychologist Neil Malamuth of UCLA, one of the foremost rape researchers in the country, emphasizes why. Malamuth had students read two sexually explicit tales, one involving graphic, consensual sex, the other detailing a rape in which the woman, despite obvious initial pain, showed some signs of response. The students were asked to rate the stories in writing, and they were also measured for penile tumescence. In the self-report, they all insisted the consensual sex was more arousing. But their penises told a different story—the response to either tale was virtually the same.

There's a caveat here, though. The rape tales tend to take two versions. Either the woman is totally resistant and miserable throughout, or there's a suggestion that she eventually responds, and warms to the act. It was the latter, the rape stories with response, that the college men found exciting. They were clearly appalled by stories of unrelenting painful rape. It turned out later, though, that scientists could manipulate that distaste in a somewhat disconcerting way. In another study by Malamuth, men listened to an audiotape describing a man with a knife who viciously rapes a woman. She hates and fights the assault from start to finish. First, a man read the story on the tape. There was no sexual response among the college students (although other tests have shown that some convicted rapists are aroused by such stories). But Malamuth found that if he had a woman tell the tale, then suddenly university students began to find the story more sexually exciting. His speculation is that this is a near knee-jerk response: a female voice talking about a subject that involved painful sex became simply about sex.

There's another very obvious difference between male and female arousal. In order to have sex—more than that, in order to produce the sperm and fertilize an egg—men have to be aroused and they have to reach orgasm to ejaculate. Women, on the other hand, in the most basic biological terms, need neither excitement nor pleasure to produce the egg. We ovulate by a biological clock; we don't start pumping up the supply in response to sexual excitement. Biology is really unfair on this point. It has designed men so that the most casual sexual encounter can produce pleasure; climax is, in a sense, merely the successful push of the pump. But climax isn't technically as vital to reproduction in females, so women don't get equal opportunity. They don't get the guarantee. Some biologists, in fact, are baffled by why women experience orgasm at all.

The scientific literature still contains all kinds of theories on the topic of female orgasm; a recently popular idea is what Randy Thornhill calls the "upsuck" theory, that the muscle contractions help suck the sperm where it needs to go. To some extent, the implication is that sex is partnership—a man who wants to see his genes continue on thus wants to ensure his partner's pleasure. But the theory is also promoted as a female counterstrategy to rape. That is, if a man forces sexual relations on a woman, her biological defense against pregnancy is to make it harder for the sperm to reach target. (We could argue that lack of lubrication would also be a form of defense.) The concept of the body working against unwanted sperm is more of an idea than a theory with any substantial evidence behind it. The best support comes from study of orangutans, one of the few other primate species in which clearly forced rape occurs. Orangutans live in a rigid harem system, and for young males, unless they reach alpha status, forcing sex may be their only chance at reproduction. Yet studies of orangutans show that such forced copulations are surprisingly unlikely to produce offspring—at less than half the rate occurring from copulations with the chosen alpha male. Researchers have speculated that female orangutans' warm response to the selected mate may improve the odds of pregnancy, although it could also be that the alpha male is something of a supermale, with very potent sperm.

There's also an opposing argument, that orgasm works against reproduction, causing vasoconstriction, or squeezing the sperm out. Donald Symons suggests that orgasm in females may occur simply as a kind of evolutionary accident. The idea is that perhaps, as sex was evolving, a remote ancestor started with a neutral structure of sorts, which developed into a penis in males and atrophied in females, becoming a clitoris, which is smaller, less necessary, less quickly aroused to pleasure and less reliable in that sense. If you believe Symons, it's possible that women have an imperfect remnant, a less-developed version of what males possess. On the other hand, how then to account for the fact that women have multiple orgasms and men don't? The hypotheses, as you see, tend to be contradictory. That tells us that scientists don't understand the biology of orgasms in females.

The one constant is that orgasm is more necessary to reproduction and, therefore, more predictable, in men. According to sex surveys—which we should probably regard with some skepticism—men average about a 90 percent or higher success rate when it comes to orgasm. Women are far behind, at some 30 and 50 percent. A recent study asking university students about their first sexual experience found that both sexes were very nervous, that men were more worried about performance, and that women tended to feel guiltier. And the pleasure ratio: 79 percent of men reported orgasm compared to 7 percent of women (taking into account that the first sexual act can be painful).

The numbers really say this: with a skilled and considerate partner, most women can thoroughly enjoy sex. But men don't need the skilled partner; they can go to bed with a nervous and unskilled virgin and still climax nearly 80 percent of the time. The same system that revs them up at the sight of a centerfold, that causes teenage boys to develop erections from the vibrations on a school bus, and that makes casual conversation a turn-on tumbles them more easily into pleasure. But, conversely, that puts more pressure on males to perform well in bed—the underlying message in Randy Thornhill's upsuck theory of orgasm.

There's some obvious tension involved in this particular difference—a frustration on either side. And science itself doesn't escape that. I was reminded of this during a debate over the elusive "G" spot—the supposed pleasure point in a vagina—which appeared on Sexnet, one of the Internet talk groups for sexuality researchers. After a long discussion of whether the spot really existed and where it was, and why most nerve endings in the vagina were near the entrance, one scientist mockingly posted the following joke:

What's the difference between a G-spot and a golf ball? A man will spend 20 minutes looking for the golf ball.

Yet, even left to themselves, women don't pursue orgasm with the apparent single-mindedness of men. Theoretically, anyone, male or female, can achieve 100 percent orgasm success through masturbation. For comparison, I'm going to use numbers from the 1994 University of Chicago's sex survey, one of the biggest and most sophisticated ever done. The Chicago researchers found that about 45 percent of married women masturbate on a regular basis, more often, actually, than single women. They proposed that the higher masturbation rate among married women reflected the fact that they are more sexually active. Overall, women with a partner they like engage in sex more often than unattached women. But for married men, the masturbation rate was still considerably higher, some 85 percent. I'm not suggesting that we should trust one survey; every sex survey done has found a similar imbalance. And the overall suggestion is that, still, men enjoy the simple mechanical aspect of pleasure more than women. Some scientists suggest that there's an underlying difference in attitude as well—that in evolutionary terms, sex is serious business for women. They don't relax into it as quickly. It's easier for men to regard sex as a kind of recreation.

It may be, as well, that a system primed to constant arousal needs more frequent relief. For instance, we know that testosterone levels are tuned to arousal; they rise in anticipation of sex, they rise after sex, they rise in response to thinking about sex. Some recent studies by the Kinsey Institute show that women with higher baseline testosterone levels tend to masturbate more often. And this, although researchers are still puzzling it out, could be no more than a comparable need to scratch a sexual itch. Since, in men, testosterone is associated with a strong sexual desire (for someone else), obviously it's not performing an identical job in women. But is there a hormone that drives sexual desire in females? There's research that links rising levels of estradiol in women with rising sexual desire. But that's very different from testosterone, which can spike up any time, day or night, if a man is attracted to a potential partner. Rising estradiol in women does not follow lust-at-first-sight. It peaks during ovulation, following the schedule of the body preparing for the possibility of pregnancy.

One of the best ways to test that connection, it turns out, is by studying oral contraceptives, because it also turns out that "the pill" is something of an experiment in behavioral biology. We tend to think of oral contraceptives as merely a slicker, more efficient version of a diaphragm. But that's simplistic. We know, in fact, that oral contraceptives such as Depo-Provera can reduce desire in men by blocking androgens. If we give women a medication that alters the natural cycle of their primary reproductive hormones, shouldn't we expect some impact on their sexual behavior as well?

Remember the sweaty T-shirt study—the work by the Swiss biologist Claus Wedekind who found that a woman finds the smell of a man sexier if his immune complex is different than hers? Wedekind believes there's an essential signal from the immune system in that smell which helps women make a healthy mate choice. Well, he also found that oral contraceptives reversed that choice pattern, almost completely. Women taking oral contraceptives were most attracted to men with similar immune systems. "The contraceptive pill seems to interfere with natural mate choice," Wedekind noted. He suggested that steroid hormones such as estradiol play a poorly understood part in the smell of attraction. By altering the natural flux of hormones, the pill apparently altered a woman's ability to detect a mate with an appropriately different immune system. The implicit message is that the normal rise of hormones with ovulation is tuned to mate selection, through senses such as smell. And the unanswered question is whether the pill has influenced a whole generation of women toward the "wrong" men.

And does the hormonal swell carry arousal as well? Do women

find men more sexually stirring at the right time of the month? A study of !Kung San tribes in Africa, which included blood sampling of hormones, found that with ovulation and rising estradiol, women reported increased lust for both husbands and lovers. The pill also provides some very suggestive—although not totally conclusive—studies along that line. In 1978, the *New England Journal of Medicine* published a study tracking two groups of women, one using oral contraceptives, the others relying on diaphragms, IUDs, and your basic collection of sperm buffers. Each of the 35 participants kept a daily log of sexual activities and of what scientists call "interest," but which is a lot more fun than that: more the way sexual fantasy drifts, unbidden, to mind, the way arousal jolts, unexpectedly, as you watch a partner in the morning, rubbing sleep away. Women who were not on the pill reported a rise in both interest and activity timed closely to ovulation. Women on the pill reported no particular change in interest or initiation.

A recent study done in Germany found that women on birth-control pills see the world as a far more platonic place than women using other forms of contraception. Researchers at the Medical University of Lubeck in Germany showed a series of images—men and women stripped to the basics—to students on the pill and off. Even if women tend to be less excited by nude pictures, it's fair to say that most of us, looking at a naked body, might let some possibilities drift across our minds, perhaps on to the lover at home. Compared to other women in the study, those on the pill were far less likely to do that and more likely to see a body as a neutral composition of muscle, bone, and skin. They were also less likely to be charmed by pictures of babies, another hint of influence on reproductive interests.

Some of the most intriguing work on oral contraceptives, estradiol, and desire has been done in primates, notably by Ron Nadler, a primatologist at the Yerkes Regional Primate Research Center. His work shows a direct connection between hormone and lust, but, even more interestingly, it also shows that, for females far more than for males, the relationship itself can override the urge—that even among the great apes, for the females an element of affection becomes a potent part of sexual attraction.

Nadler's first studies were done with orangutans, the big apes that live in such a rigid social hierarchy. Working with captive orangs, Nadler found that males were basically always turned on. Introduce a female into his cage and you could make money betting that sex would happen that very day. But what happened if you gave the female some power? Nadler caged the orangutans side by side, so that they could see each other. The separate cages were connected by a door, but it was too small for the bulky males to pass through. Only the slighter females could use the passage. As it turned out, most of the time a female orangutan was content to admire the male through the bars, and to leave him there. The females used the door only when they were ovulating. And, at that point, you could make money betting that sex would happen that very day.

Nadler then decided to explore that hormonal relationship further, by studying chimpanzees and by testing them on and off oral contraceptives. As with the orangutans, he first put males and females together in a single cage, where the bigger, stronger, and meaner males tended to dominate and the females to ease into a safe and submissive posture. These were chimpanzees that knew each other well and were clearly accustomed to each other's company. The researchers then separated them as they had the orangutans: side by side in a divided cage. The male could only reach the female if she pushed a lever to open the connecting door. The lever was only on her side.

"Male chimpanzees can dominate female chimpanzees," Nadler says. "Wild and captive male chimps have been observed to intimidate female chimpanzees to copulate when the females couldn't avoid them or solicit support from others in their social group. We used the restricted access test to prevent the male from intimidating the female and thereby to better evaluate the females' motivation to mate." And then he put some females on the pill and compared them with those who were not. If female orangutans received the males only when estradiol prompted a mating urge, what would happen when oral contraceptives suppressed estradiol in chimpanzees? Would they be content simply to watch the males into infinity through bars?

There were nine chimpanzee pairs in the study. Once females went on the pill, sexual relations declined steadily for seven of them; the door stayed shut more often. With one of the remaining pairs—who apparently had a terrific relationship, sexual and companionable—the oral contraceptive made no real difference at all. They threw open the door and mated with undiminished enthusiasm. And with the other, where the male had aggressively intimidated the female, mating habits were also basically undisturbed. In other words, the relationship—good or bad—could override any effect of estradiol suppression.

As Nadler pointed out, the critical difference was not the oral contraceptive itself, but female control. On the pill, when the animals were housed together, sex continued according to the males' wishes. But when the female operated the door, the pace changed unless there was a strong relationship. Among the chimpanzees, male dominance and compatibility were the strongest predictors that the door would stay open.

Does this transfer to humans? Of course it does. Do women have sex with men they like, even if they aren't blazing with lust? Do they sleep with men because they're afraid of them? Those are rhetorical questions, obviously. The message Nadler reads in the chimpanzee studies is twofold: oral contraception does suppress desire, and that result may—depending on the relationship—alter sexual activity. Since humans are less hormonally hardwired than chimpanzees, he says, the influence of relationship should be even stronger. There may be some women who are unusually sensitive to estradiol effects and who may find that the pill changes a relationship. A woman who doesn't like her current partner much anyway may be even less enthralled while on the pill. Oral contraceptives could thus possibly make a troubled relationship worse.

But it appears they're less likely to have much influence on a good one.

This, too, you might expect in the evolutionary biology profile of a female being most attracted to a good partner. The study of estradiol suggests that it prompts desire in terms of reproduction and reproductive timing, not that it prompts a woman to get excited about every man she sees. In his influential and occasionally very funny book, *The Evolution of Human Sexuality*, Donald Symons comments that "the sexually insatiable woman is to be found primarily, if not exclusively, in the ideology of feminism, the hopes of boys, and the fears of men."

In fact, oral contraceptives seem almost paradoxical. They may suppress desire, but they may also liberate women to be sexually aggressive. By virtually eliminating the fear that every sleep-over could produce a child, they also eliminate the restrictions imposed by choosiness. A woman doesn't have to worry that the one-night stand may not be the ideal father/partner. She has that under control, she's suddenly free to be promiscuous—if she chooses—in that carefree "male" way. (That's not to say that men aren't also responsible for contraception, only that female contraceptives, so far, are the most effective on the market, partly because they can be controlled by the partner with the most at stake in avoiding an unwanted pregnancy.) Kim Wallen points out that fear of pregnancy is also an effective check on desire, so that even if estradiol did drive sexual interest, many women might still choose abstinence over risk.

Wallen proposes that steroid hormones are a primary biological driver behind desire—so that women can be equally as sexually charged as men, but not as continuously. This is a straight contrast between estradiol, peaking on its monthly cycle, and testosterone, charging up at the sight of a centerfold. And Wallen raises the point with a tone of envy in his voice: "Women get a chance to be really sexually interested and then they get to think about something else for weeks. Men don't know what that's like."

The trick is not to exaggerate this portrait of differing male and female desire beyond what's real. Clearly, as both Wallen's and Nadler's work emphasize, females, human and otherwise, can and do feel intense sexual desire. And they also can and do pursue males with unmistakable aggression. The most obvious examples occur in polygamous monkey societies such as rhesus macaques. Rhesus society, as I mentioned earlier, is female-dominated, thanks mostly to the way the female monkeys band together in mutual support. As a group they are intimidating enough that during the mating season the males wait for a clear signal of female interest. A male who hassles an uninterested female is likely to find himself under attack by a screaming, and dangerous, band of her allies. So, in the mating season, the females sometimes pursue the males for hours, sitting nearby, grooming and hinting until the male relaxes into approaching her. And frequently, the females will mate with one male, and then try out another.

Scientists sometimes call this taste for sexual novelty the Coolidge Effect. It takes its name from a anecdote told about former U.S. President Calvin Coolidge. The story goes like this: One sunny day, the President and Mrs. Coolidge were visiting a government farm. They separated, each to tour the farm at a chosen pace. When Mrs. Coolidge passed the chicken pens, she asked the guide if the rooster copulated more than once each day. "Dozens of times," was the reply. "Please tell that to the President," Mrs. Coolidge requested. When the President arrived at the pen, he was also told about the rooster. Coolidge asked: "Same hen every time?" The guide shook his head, saying no, the rooster approached a different hen every time. The President nodded slowly, then said, "Tell that to Mrs. Coolidge."

In cattle and sheep, the Coolidge Effect is so strong that you can wear a male into the ground, almost literally, by changing partners. Put a new cow in with a weary bull and he charges up all over again. Researchers have tried fooling rams, disguising a familiar ewe by putting a bag over her head, only to have the male amble off in boredom. If they bring in a new female with a bag on her head, the ram reignites, so to speak. The Coolidge Effect is often cited to make the point that males love sexual novelty, as indirect evidence of the masculine drive to engage multiple partners.

But, as Cornell University anthropologist Meredith Small points out, males aren't the only ones fascinated by the siren call of new genes. Rhesus macaque females risk the wrath of their group's alpha male to try other partners, and, even though the boss male will try to muscle apart such pairings, the females will continue their pursuit. The same is true with other monkeys. Barbary macaque females "float" from male to male; Japanese macaques have as many as nine partners during estrus. Small insists that there are valid reasons why females might benefit from more than one partner. Perhaps they want to evaluate a male, she says, to test him out. Or perhaps hanging about with one male is a waste of energy too. After all, when a single male ejaculates too frequently, his sperm count goes down. Males need breaks. "This vacation might be frustrating to a female monkey in heat," Small points out.

Small and other feminist scholars have been determined to dispel the notion that females are designed to be sexually passive. Often, even in monogamous species, the females are more likely to experiment on the side than the males. That was Gubernick's finding, of course, in California mice. And the same has been found in other species such as titi monkeys. Bill Mason and Sally Mendoza have done experiments in which they introduce a new female or a new male to a longtime titi monkey couple. The males are usually indifferent. But the females are often interested, and Mason has watched males physically try to hold the females back from the stranger. It's exactly because females are interested in multiple partners that monogamy often predicts sexual jealousy in males. Thomas Insel, of Yerkes, notes that as soon as a gentle little prairie vole mates, his personality changes to become more suspicious, more aggressive with other voles, more ready to bite on sight. If a male makes a serious investment in a single female and their offspring, then her infidelity becomes serious—he may end up raising someone else's children, a genetic catastrophe of sorts. Put simply, male sexual jealousy is part of monogamy because females are interested in sexual variety—or at least in a kind of genetic comparison shopping.

That awareness has gradually replaced a long-standing notion in science that females were basically empty vessels, waiting to be filled. Why have we given up the old, inaccurate stereotype of the passive female? Some of the credit goes to Small and a host of other top-flight female scientists. As women entered biology—once, like so much of science, a male-only profession—they influenced the perspective. They saw the world from a different angle; they helped provide a more sophisticated understanding. Consider the early twentieth century description of human females as "continuously receptive" to sex. No one, man or woman, uses that term now. They'd be ridiculed. But it's also worth remembering that twentieth century male scientists were already moving away from such Victorian misconceptions. As early as the 1940s, the noted primatologist Charles Carpenter reported that monkey females actively court sex. And Frank Beach, one of the best-known comparative biologists, ripped apart the notion of nonstop female receptivity. Beach wrote a compelling research article in 1976 which focused attention on female sexual initiation. He commented that "any male who entertains this illusion must be either a very old man with a short memory or a very young man due for bitter disappointment."

Did the vessel notion come out of some antiquated world view of women's place, so that researchers didn't watch animals objectively, screening them through the sexual politics of the day? That's undoubtedly at least partly true. On the other hand, it's a mistake to let our current sexual politics, emphasizing equality, blind us. Objectively, it's still a lot easier to find females patiently accommodating excited males than vice versa.

One of the best examples of this comes from African spotted hyenas, a species which in many ways reverses sexual stereotypes. Hyenas are the natural world's demonstration, in a sense, of how far androgens will push a female body toward masculinity. In species such as our own, although some levels of testosterone circulate in the mother's blood, the placenta blocks it and keeps it away from the fetuses. Male fetuses are exposed to testosterone because their own bodies make it and females (unless they are near a developing male, such as with a boy-girl twin pair) are essentially androgen-pure before birth.

The opposite case exists with hyenas. During pregnancy, androstenedione—a precursor of testosterone—builds in the mother's blood; her T-levels rise tenfold, and the placenta itself pumps testosterone directly to the young. In hyena males, this adds up to normal masculinity. But in females, it's transforming. Female hyenas are bulkier than males, more aggressive from birth. They dominate the hyena social structure, and their reproductive package itself looks male. The females sport a dangling, tubelike structure that looks like a penis to anyone who doesn't hang out with hyenas. The best hyena colony in the United States is tucked away in the dry grasses of the Berkeley hills above the University of California campus, where neuroscientists Stephen Glickman and Lawrence Frank have been hanging out with hyenas for quite a few years. Yet despite their familiarity, it was almost a decade before Glickman realized that you could visually distinguish a male penis (slightly more pointed at the end) from a female tube (slightly more blunt).

For all of their "male" attributes, female hyenas are dedicated mothers. They are heterosexual and they allow the males to do the courting. It's one of the aspects that fascinates and puzzles Glickman—that the females can be so radically transformed in shape and aggressive behavior and yet remain dedicated and nurturing mothers who build a strong social network. Female hyenas are born vicious, biting their siblings as soon as they tumble out. But they stand passive for sex. In fact, they'd never get pregnant if they didn't stand passive. Hyena sex requires an inordinate amount of precision. The female first retracts her dangling tube into her body. Then the male has to insert himself into it. "She has to be very patient," Glickman says, with what strikes me as understatement at its best. Okay, I admit it—I listen to Glickman describing hyena sex, with the male poking and prodding and missing, and poking, and missing—and it just cracks me up. I can't help it. In my mental picture of this, the female is rolling her eyes, or maybe thinking about something else, like whether they need to pull down an extra antelope for dinner. But the fact is, if you're a scientist out there in the Serengeti, lucky enough to be watching hyena sex, it doesn't matter how aggressive the female is normally, she's going to look like a passive little babe—the male's the one actively poking around.

There's an interesting comparison to this in pigs. A sow is normally as bristle-minded and tough a female animal as you might happen to run across. The intelligent male pig avoids sows with great care, regarding them—rightly—as mean and ornery. So sex among pigs turns out to need a kind of male protection system built in. As a result, when a female pig is ovulating, a good whiff of boar sweat will cause her to stop in place. The chemistry is so potent that the male doesn't actually have to be there, just the sweat. If researchers rub a little boar sweat under the nose of an ovulating sow, she turns into a living statue and they can comfortably climb on her back until the effect wears off. When it does wear off, she dumps them flat.

Kim Wallen makes a smooth comparison between these immobilized pigs and some similar behavior in rodents. In female rats, the vaginal opening normally points toward the ground. A male rat can get as excited as he wants, but he can't get in unless the female raises her body, a process called lordosis, which elevates and rotates the vagina, making it accessible. And a female rat simply won't do that unless she gets a signal from the ovarian hormones at ovulation time. A male rat can be a walking bundle of sexual nerves, but until her hormones say yes, that's his problem.

"Lordosis has almost nothing to do with sexual behavior in humans," agrees Marc Breedlove of Berkeley. "It's a reflex," one controlled by estradiol. If Breedlove injects a female rat with an ovulation-sized dose of estradiol and waits about six hours, he can induce the female rat to present for mounting just by touching her genitals with a finger. "She's all freaked out about my being so close but the hormone makes her give the right response," he says.

As with the hyenas, there's little evidence that the female finds any pleasure in mating. With rats, in fact, it's likely that she doesn't. "Male rats have a nasty penis," Breedlove observes; the outer skin is actually barbed. In rat sex, the barbs do seem to provide some stimulation necessary for getting pregnant. But a little goes a long way, and if a male rat doesn't perform quickly and efficiently, the female wants him off, off, off. She'll bite and push at him. Still, the estradiol compulsion is strong enough that female rats will approach males during ovulation, pursuing sexual contact. Does the female rat get any pleasure at all? It's difficult to say. She seems to have some protection against pain, anyway. One of the scientific findings is that lordosis induces a natural anesthetic response. Maybe the anesthetic response lasts only so long and then, ouch!

This is desire indeed—an urge to mate that overrides a near-guaranteed painful experience. It's not equal to the suicidal heroics of male spiders, such as Australian redbacks or American black widows, who are routinely killed by the larger females as part of the mating process. But how many women would feel any desire at all for sex if it always involved barbs and hooks? Although historically, scientists have looked at the motionless pigs as examples of passivity, it's clear that they also represent an extreme of a hormonally driven mating urge.

The rats, the pigs, the hyenas, all emphasize both the potency and the specific timing of estradiol in driving desire (timing meaning ovulation). It's more difficult to read the connection in humans. We don't advertise ovulation visibly; as Nadler points out, our emotional desires may control our physical desires; and the way biology and behavior interact is a lot more complicated for us than for rats. Meredith Small suggests, for instance, that "women are so culturally bound to be passive sexual partners that it's almost impossible to know what the truth is."

There's one more animal comparison worth making at this

point. If human females were unfettered by all the rules that say that "good girls don't," then would they be more sexually open and promiscuous? Would our society be more like bonobos—once called pygmy chimpanzees? Bonobos are considered to be a hair's breadth more genetically like humans than are common chimps. They stand a little more upright, their features are a little finer, and some researchers believe that they are a little smarter. Language studies by Sue Savage-Rumbaugh at Georgia State University have produced some spectacular results by focusing on bonobos rather than common chimps. In those studies, the chimps use a computer keyboard to pick geometric symbols that represent words. They also string them into simple sentences: "I want food; where's my toy?" Savage-Rumbaugh has expressed the opinion that bonobos are swifter than their common cousins at picking up the complex, abstract ideas that lie behind language.

Bonobos are also, at least by Western standards, spectacularly promiscuous. Let's talk continuously receptive here, and I mean males *and* females. Bonobos weave sex into all aspects of their lives: males mount females; sometimes females sit astride males; females rub other female genitals for fun; males do the same; juvenile monkeys suck each other's genitals; all bonobos French-kiss; and sex between two bonobos is considered an open invitation for others to amble up and insert an additional finger or toe into an orifice. Sex is for reproduction, for fun, for gratitude, for friendship—exuberantly shared and enjoyed by all.

Would we resemble bonobo chimps if we weren't confined by the rules for correct sexual behavior that guide our society? We're so far from that point that you might as well call the question a fantasy. Frans de Waal of Emory, who's done some of the most detailed studies of bonobo sexual behavior, points out that some people can't even stand to see bonobo behavior. Some of the early graphic photographs of bonobo sex—and this is pretty creative stuff, from which we could all learn a lot—were published on magazine covers in Europe but not in the United States.

From what I'm going to call a feminist position—toward complete equality—bonobos represent a kind of sexual ideal. Meredith Small describes watching films of bonobos with a kind of wistfulness: "It looks so human—actually better, and even more fun than human sex." Imagining humans as bonobos, though, illustrates how difficult it is to separate the influence of culture from biology. How easily would we change our society so that we cheerfully exchanged sex for simple favors—a meal, a lift in a car—and encouraged our children to do the same? If we had the biology of bonobos, would our culture feature free-for-all sex? Or have we used culture to repress the biological potential within, in this case, for open sexuality?

And if so, why? Could sexual repression in human culture be based on our unique ability to guess at complex cause and effect? Perhaps early hominids learned that those who engage in unrestricted sexual intercourse with multiple partners suffer consequences in the form of disease. Could ancient societal rules have grown out of such recognition? If we argue that human society is male-dominated, we might suggest that rules would first protect men by controlling "sexually indiscriminate" females who might be considered dangerous—thus, the good-bad connection, a link to what we now call morality. Could tribes with more restrictive rules about sex have survived better in times of rampant disease? Data about prehistoric venereal diseases are nonexistent, so we can only speculate.

It's possible, even likely, that human sexual morality is a lot more basic than that. Many biologists argue that our culture is based on our distinct biology, but—following Randy Thornhill's point—natural doesn't always mean right. Consider again the underlying tensions in the male-female relationship between Bill Rice's fruit flies. Extend that to humans and you could argue that—irrespective of and probably long predating any human thought about personal or tribal well-being—basic biology would predict that men would try to use culture to control women, and vice versa, and that the resulting culture would reflect who was winning the sexual arms race at the time.

Rape is not a part of bonobo society. Considering how free and easy sex is, why would it be? And how do you explain that we—with all our righteousness about sexual morality and privacy—have been unable to eliminate forced sex, sometimes vicious forced sex, from our behavioral repertoire? In the United States, reported rapes average about 70 per 100,000 women every year. That's reported, successful rape; police reports indicate that attempted rape is almost three times that number. In cold statistical terms, those numbers remind us of one very important point—this is not a majority crime. Most men are not and will never be rapists. A majority of women are not attacked. But the numbers are high enough and persistent enough to tell us that rape, even in a minority sense, is very difficult to make go away.

On this point, psychologist Neil Malamuth, of UCLA, emphasizes the difficulty of separating biology from culture: "The common thread is that men perpetrate coercive sex against women and typically not vice versa," Malamuth says. He points out that it's known, in times of war, that rape rates skyrocket in some potent mix of fighting, power, and sex. At least, a lot more men rape a lot more women. Adds Malamuth, "In today's [more equalized] culture, it's feasible that a woman in Bosnia could put a gun to a man's head and say, 'Give me oral sex or I'll kill you.' But I haven't heard of that happening. It's not just a matter of male and female physiology but of psychology. The minds of women don't seem to have the similar capacity for coercive sex."

Malamuth is slightly bemused by how much people seem to hate the idea that the male sex drive may include a tendency toward force: "People seem to be able to accept in our ancestors that the capacity to kill, under some circumstances, might be naturally selected for. But if you argue that's also true of the capacity to rape, the reaction is very quick and very negative."

And yet total revulsion against rape is actually fairly recent. Susan Brownmiller, reviewing its long history, concluded that past attitudes derive from a culturally imbedded belief that men have a right to dominate and possess women. The word "rape" comes from the Latin word "rapere," which refers to seizing of property. That notion of women as property, of men having a right to sex on demand, is best reflected in the fact that spousal rape was an unknown crime until recently. The practice was considered a private matter, part of the marriage. Oregon was the first state to outlaw marriage as defense against a charge of rape, and that was in the late 1970s. California adopted such a law in 1979, over serious protest from some state senators, one of whom complained, "If you can't rape your wife, who can you rape?" We haven't left those attitudes as far behind as most of us would like to believe. In 1995, a group of Cornell University freshmen sent out a mass e-mail, listing 75 reasons why women shouldn't have freedom of speech. Reason number 38: "If she can't speak, she can't cry rape."

Robert Prentky of the John J. Peters Institute in Philadelphia, who studies violent rape, says our society remains fixed on the notion of women as sex objects—emphasis on object: "The messages of our entertainment media, for instance, are clearly that women should be treated as sex objects and that doing so is a socially approved vehicle for demonstrating one's virility. How do we sell cigarettes and beer and cars? We place them in the hands of barely clad women. Thus, it is impossible to insulate an offender from the messages that reinforce the attitudes and behaviors that [psychological] treatment sought to extinguish." In fact, Prentky suggests this is one reason that rapists reoffend much faster than child molesters. Our culture, including the entertainment media, condemns child abuse; there's no reinforcement for the behavior.

Rape runs unusually high in the United States: three times higher than in England, West Germany, Sweden, and Denmark; five to ten times higher than in France, Holland, Belgium, and Japan; and about two times higher than in Canada.

Those numbers suggest that if there is an inclination to rape, a more violent culture aggravates it. There are some interesting studies along that line, connecting violence in the media and violent behavior. It turns out that if men watch violent pornography and then begin to flirt with a woman, they are more likely to be furiously insulted by any rejection and more likely to respond aggressively to it. In another study, scientists found that men who listened to an audiotape of a grisly killing had a much stronger sexual response to photos of a pretty woman than a comparison group of men who had just been relaxing. In a last example, men were surveyed on their attitudes toward slasher films and whether they were demeaning to women. Many questioned said they did find the movies sexist. They then watched ten hours of such blood-spilling, cut-them-up films. At the end of the viewing session, they were surveyed again. The men no longer perceived the movies as quite so offensive and antiwoman.

Prentky was particularly outraged by the 1994 movie *Kika*, from Spanish filmmaker Pedro Almodovar, which included a rape scene played for laughs—the rapist being so incompetent that the woman begins to nod off. "Imagine a rapist watching that movie and listening to people giggling away at something he was sent to prison for," Prentky says. "We are just constantly besieged by messages that endorse sexual aggression. Yes, you can argue that a large percentage of the population is not rape-prone. But how do you separate out the reinforcing messages of sexism which are so ingrained in our society?" And was it our biology that built the attitudes in the first place? "I don't know that you can ever separate the two," Prentky says.

"When you talk about culture, you're talking about something that humans created; it's a product of the human mind and mechanisms of that mind," says Malamuth. "Societies vary enormously in rape-proneness and rape-discouraging. Our society may send mixed messages about the appropriateness of certain behaviors in this context. With the emergence of the feminist movement, there are more messages that discourage it."

Say, then, that for a long time we've had a culture relatively tolerant of rape. And that now we're less tolerant of it—largely because women have organized and fought to make everyone see that rape is intolerably wrong. No state senator who wanted to stay in office these days would complain about denying a man his right to rape his wife. I'm not arguing that men rape simply because culture confuses them about whether it's right or wrong; rape is far more individual than that. If you look at studies that try to profile the rapist, you find, as with all human studies, that to some extent each man stands alone—he's a singular mixture that may include troubled home, abusive father, insecurity, and anger, and a belief that women should be controlled. He may drink; the association between too much alcohol and rape is extraordinarily high. It may be that media messages reinforce some of his attitudes. And it may be, too, that as, or if, the culture becomes less accommodating to woman-as-object, he'll get less reinforcement.

And it's in that context that I want to raise once again the fruit fly question. What I wonder is: Are we in the midst of a cultural sea change in which rape, finally, is becoming unacceptable, however painfully slow that change may seem? And if so, does it represent a power shift in the basic tug-of-war between men and women? The rise of the feminist movement, the organized opposition to rape and spousal abuse, the new spate of laws curbing men's sexual behavior—do these all represent a societal-evolutionary counterstrategy? And if they do, where will the strategy take us?

Through political force, cultural disapproval, the protection offered by the pill, and even the active means of abortion in cases of rape-caused pregnancy, is it possible that we could begin to weed out genes that reinforce the tendency toward rape among human males? Would other behavioral changes related to tempering sexual overaggression and/or violence follow? If human culture is indeed an aspect of nature, flowing from human biology, then we can imagine a feedback loop in which biology—as in matters of mate selection—also flows from cultural influences. And if we could carefully direct cultural change, we might, gradually, influence evolutionary changes in our species. We could do so without even realizing it.

Across species, females evolve protections against the dangerous side of males. Fruit fly females may counter toxic male chemistry with biochemical antidotes. Female sharks have thicker skin than male sharks, buffering them against bites during copulation. Among African swallowtail butterflies, there are two different female shapes, one looking more like the gaudy male. These male mimics, scientists have found, are hassled less by real males. And when they are in the mood for mating, they release a pheromone that summons the ever-randy real males. Monkeys, such as rhesus macaques, often form tight female-female bonds to protect themselves against males, and an individual female can rapidly summon reinforcements against an overaggressive suitor. To the dismay and puzzlement of feminists, women seldom network as effectively as rhesus females do, uncompromisingly together against males when needed. Women have in fact been among those who bought into the historical good-girl, bad-girl dichotomy, into the "she asked for it" defense against an accusation of rape. Women have even accepted the "she meant yes when she said no" line. We haven't always been our own best allies. That, I hope, is changing.

Whether such changes belong to a sexual power shift, or whether it's a short-term trend or an evolutionary one—these are interesting and even important questions. But they are not the kinds of questions that need paralyze us. We can always try to improve where we are. However complex and contradictory male and female sexuality may seem, most of us agree on at least one aspect of the debate: we all have the right to choose our own bedmates. And we all—men and women—have the right to choose, on any given night, no bedmate at all.

Chapter Nine

ONCE DIVIDED

Power and Gender Biology

Here's what I want to know: If men and women are equal—divided mainly by a few structural nitpicks in the brain and some invisible hormonal tides—why don't we have equal access to the world?

Can biology explain this gender difference? Science offers all kinds of interesting explanations and ideas about the way men and women behave, especially toward each other, but does it ultimately explain why men have the power?

Not every man today holds an enviable position of power. Not every woman is suppressed. Many couples, most of those I know, anyway, strive for some sort of equality in the home. Businesses make real efforts (along with paying a lot of lip service) to achieve and enforce some kind of sexual equality in the workplace.

Still, overall control over big decisions—the ones governing the future of a major corporation, of a major city, of the United States, of Western society, of the human race at large—remains largely in the hands of men. Women do rise to positions of real power these days, but often only by adopting the traditional values of the men who have come before them. Their success, or lack of it, is measured by the standards of a man's world.

What led humankind—that is, men—to build and maintain a civilization so lopsided? How did men get to set the rules of the power games? How did we come into this century so divided that British philosopher Bertrand Russell complained, some 50 years ago, that women were invisible in world affairs, that "they have been kept artificially stupid—and therefore uninteresting." And Russell was progressive for his time; at least he proposed that women had potential. It wasn't so much earlier that scientists were suggesting that women could be classed with gorillas in terms of intellect.

There's nothing in what we've learned about brain structure, test scores, sexual drive, or parental behavior to say that one sex should be considered more deserving or more valuable than the other. I'm hardly alone—at least in this country, in the 1990s—in that opinion. I'm practically surrounded by bookstores full of self-help books explaining, in the most politically correct way, that men and women approach the world differently, that we need to understand those differences. But the self-help books don't answer my question. They don't tell me why we haven't inherited a system of equality based on mutual respect. So I wonder if the answer to that question lies in the still-shrouded mysteries of our evolutionary past.

In a brilliantly edgy analysis called "The Man Who Mistook His Wife for a Chattel," evolutionary biologists Margo Wilson and Martin Daly point out that we are barely removed, even in Western society, from a time when wives were property. The term "rule of thumb," for instance, derives from an eighteenth century judicial ruling in England, saying a man could physically discipline a disobedient wife, but it was not all right to kill her; he was legally allowed to use a rod no thicker than his thumb. Even in the nineteenth century, Englishmen conducted wife sales from the same blocks used to sell cattle. Girls were led to auction with halters around their necks. Until a 1973 ruling made it illegal, an Englishman could legally hold his wife captive if she sought to leave the marriage.

Today, of course, most countries consider imprisoning and beating a wife to be illegal spousal abuse. But a true balance of power within marriage, at least as signified by a true balance of labor, seems yet a long way off. In a 1994 study, University of Michigan researcher Marjorie Starrels found that in dual-career households, women on average do three times the housework that men do. They work a minimum of 15 hours longer each week on "traditional domestic chores." Even more to the point—my point, anyway—other studies find that if the woman is employed and the man is not, he still does less housework than she does. Men interviewed for the study said they would find it demeaning to have to dust and clean while they were out of work. It would be like being a wife (apparently, an awful fate), taking the subservient position in the relationship. Overall, the researchers found that 21 percent of men who earn more than wives share housework, 30 percent of those with comparable earnings share domestic chores, and those who earned less than their wives showed the greatest reluctance to clean house. This is just one study, of course, and it partly reflects class values. We can all think of couples who share better than this. But the overall conclusion is telling: "If men lose power over women in one sphere, they make up for it in another," the report said.

In that context—and certainly this isn't only true in marriages—money equals power. There are countless studies showing that men make more money than women, are paid more for the same work, and hold more of the high-paying jobs. For instance, of those in the United States making the 1996 minimum wage of $4.25 an hour, almost two-thirds were women. But, in a curious way, the fact that men make more money doesn't always mean they have more money power. Focused as I am on the bottom line—my paycheck—I raised the pay differential question with Bobbi Low, professor of anthropology at the University of Michigan, who studies the male-female power balance. She suggested that the checkbook is only part of the picture—that women have financial control beyond their own pay. They often do much of the family purchasing—gifts, groceries, clothes—and so the bottom line tends to underestimate women's power in this area.

In terms of political power, though, Low's work clearly shows

—no surprise here—that men dominate. "Despite notable exceptions—Margaret Thatcher, Indira Gandhi, and others—politics in the developed world is largely a man's game," she notes. In a survey done some 20 years ago—the most comprehensive on record—93 societies were compared. Of those, 65 had only male political leaders; seven allowed both men and women into politics, although men were higher-ranked and more numerous; two regarded women as equal; and in 20 cases, the leaders refused to answer questions, "which I suspect means that women's power is slightly overrepresented in this sample, since a major reason for not reporting relative status is the nonexistence of women as leaders," Low adds.

There's one more way to look at the power differential. As a culture, we can make all kinds of sweeping generalizations about what power is—knowledge, money, influence, control, all of the above. In terms of gender differences, though, we can also understand power by recognizing what it is to be powerless. And that, I think, cannot be told through statistics and surveys. To understand lack of power, you have to understand what it means to be a person without any. The example that stands out for me is the case of a woman—a very young woman, when the story begins—named Carrie Buck. She was born in 1908 in Virginia. But it was during that 1920s that she became—unwittingly—a symbol for a shameful time in American history.

In 1923, Carrie Buck was a 16-year-old poor Appalachian girl, living with foster parents. Photos of her show a rounded, unsmiling, quiet face, a tumble of curly dark hair, and a straight, thin body in a neat print dress. I can't put myself in her place, not precisely. I grew up in the South too, but in cities—not in the green mountains of Virginia and not poor and not alone. Still, I imagine her dreaming the dreams that most 16-year-old girls do, of a home of her own, a good job, someone to share the home with, a family, maybe. Not fantastical dreams—although she may have had those too—but hopes of a comfortable, happy life.

That year, though, she was raped by a young man she knew, a cousin of her foster mother. As a result, at age 17, she had a baby daughter. Carrie named the little girl Vivian. They were living, at the time, in Charlottesville, Virginia, but afterward, really because of the baby, she ended up in Lynchburg, Virginia, best known today as a headquarters for the religious right and for Babcock and Wilcox, maker of nuclear reactors. Lynchburg, as it turned out, was a terrible place to be if you were an impoverished teenage girl with an illegitimate baby daughter. She was put into the Virginia Colony for the Feeble-Minded there (where the state had decided to lock away its "undesirables," including mothers of illegitimate children).

This was no isolated backwater decision; the whole country was moving that way. There was, in fact, an American Association for the Study of Feeble-Mindedness. The association promoted the idea, described by its president, Walter Fernald, in an 1893 speech, that "a large proportion of our criminals, inebriates, and prostitutes are really congenital imbeciles."

Fernald—and many others like him—believed that females in particular were most easily led into evil. After all, who had the tiniest brains, the smallest ability for rational thought? And because of "the potential for such girls to become irresponsible sources of corruption and debauchery," Fernald recommended that they be kept in custody through their childbearing years, imprisoned until the apparent sanity of menopause descended upon them. That was one strike against Carrie Buck but, unexpectedly, there was another.

The IQ test was roaring into vogue. The idea that one could measure intelligence, quantify it like salt in the sea or water in the clouds, was enormously appealing to scientists. Originally, in the late 1800s, researchers began developing IQ-type tests to try to diagnose schoolchildren who might need extra help. But researchers became increasingly fascinated by the idea of testing adults, measuring them in terms of intellect. At the same time, the eugenics movement was gaining force in this country—an organization of individuals who believed that people could be neatly defined by their genes, that genetics predicted intelligence and imbecility, goodness and evil, perhaps even wealth and poverty. There's a very either-or aspect to this concept: either born smart or born terminally stupid. The eugenics movement, though, argued for weeding out the terminally stupid, sterilizing them so that their bad genes could be purified away, thus reducing the "menace of the feebleminded, the rising tide of incompetence."

Consider all this together—the low opinion of women, the belief that intelligence was inherited, that IQ tests could measure intelligence, and that stupid and promiscuous people should be sterilized—and you can almost predict what would happen to Carrie Buck. She was, after all, at the Virginia Colony for the Feeble-Minded, whose board had issued a public warning that feeble-minded women tended to breed more rapidly than "normal stock." Eight years before her arrival, the state had authorized involuntary sterilization. But the law had not been put into practice. Its proponents were waiting for a test case.

They picked Carrie Buck, who was conveniently young and naive. The directors tested Carrie and her mother, also institutionalized, with the new IQ tests. They both scored badly. The baby girl couldn't be tested, of course, but in court, a state social worker testified that the baby was sluggish and acted as if she could be stupid. The Virginia lawyers argued that Carrie was the perfect case for sterilization—daughter of lowlife, mother of lowlife, and capable of adding more lowlifes to the state's population.

They wrote up the order for sterilization. Then the state found a lawyer to deliberately challenge the order and take it to the Supreme Court. There, at the highest levels of government, the notion of weeding out "bad" genes found complete support. The high court upheld the Virginia law in 1927, when Carrie Buck was twenty and Vivian was three. The famed justice Oliver Wendell Holmes wrote a ringing opinion that declared, in part: "It is better for all the world, if instead of waiting to execute degenerate offspring for crime, or to let them starve for their imbecility, society can prevent those who are manifestly unfit from continuing their kind. . . . The principle that sustains compulsory vaccination is broad enough to cover cutting the fallopian tubes." In the case of Carrie Buck, he wrote, "Three generations of imbeciles are enough."

And so the state of Virginia essentially took up the knife and made sure that Carrie Buck would never have another child. Later, she left the Colony and worked as a maid; she eventually married. She always worked hard. As far as anyone knows, Carrie Buck was never a promiscuous woman. That analysis appears to be a mistake by Justice Holmes and the other improvers of humankind.

There's another, even more troubling twist to the story, though. Remember that those early IQ tests were experimental. Even today, with much-refined testing methods, people argue over what IQ performance really measures. As a middle-aged woman, Carrie Buck was retested with the improved tests of that time. She measured as average—not a genius, but not feeble-minded either. And Vivian? We'll never know what her IQ was because she died at age eight, of an intestinal infection. We do know that she was a B student because Paul Lombardo, a lawyer at the University of Virginia who studies the eugenics movement, found her report card. Once, not too long before she died, Vivian made the honor roll.

It's in those numbers—far beyond the marred life of poor, lost Carrie Buck—that you can read the power balance. Lombardo directs the Center for Mental Health Law at the University of Virginia. He points out that if a program is really trying to control population, the objective way to do so is to sterilize more men. As we all know, a single man can produce a lot more children (in this case, theoretically, idiot children) than a woman. The numbers suggest, thus, that the programs were gunning for women.

The operation for men was essentially a vasectomy. For women, the procedure was similar to a tubal ligation and, Lombardo notes, because it was more invasive and "because the operation required abdominal exposure, the risks for women were much higher than for men. Shock, postoperative pain, infection, and other complications made the mortality rate run at about 1 per 1,000 well into the 1940s, relatively high for a procedure that was undesired and nontherapeutic for the patient." Further, he noted, many of the prostitutes were sterilized "even after a diagnosis of chronic venereal disease or pelvic inflammatory disease, both of which usually had the effect of rendering the patient sterile without any need of further surgery. That makes it even more curious that more women were operated on than men."

Except, of course, that men were making these decisions and they decided that women were the greater evil. I don't think it's coincidence that a woman was the Supreme Court test case. I wonder if she understood the game, if she figured out how she got on the gameboard in the first place. I can't ask her because she died in 1983, still poor and, as you might imagine, alone. Lombardo did talk to her once. "When I finally found Carrie, she was living in a nursing home in a very tranquil rural setting. She was quite frail and died only a few weeks later. No one who met her later in life has reported any rage or anger from her. Perhaps 60-odd years had tempered her recollections, or at least her feelings."

Lombardo and I both respond strongly to her story, although differently. He reacts with "aggressive contempt" for the plotters, those men who obviously never saw Carrie Buck as a human being on their level or even as a human being at all. I find myself furious on her behalf. She's a lost duckling to me, out there among the foxes. I wish there'd been someone, at the right time, to truly defend her.

Many feminists, many people of all kinds, fiercely resist letting science define who we are. Their mistrust is built on stories like Carrie Buck's. They know the excesses of our own eugenics movement, and the far greater excesses of Hitler's. It was a good three decades after World War II before many people would even discuss the idea again that biology counted in behavior. There are plenty who still don't want to take up the topic. It's dangerous. It was used too well, too wickedly, to hurt and kill. And it still carries that potential.

I've talked to scientists who expressed both bafflement and frustration over women's resistance to the notion of biological gender differences. Times have changed, they argue, and science is clean again. Why be opposed even to considering the ideas, when they seem, today, so full of promise and logic? Even as it's an obvious lie to deny that a Carrie Buck is shaped by environment—by poverty, by lack of education, by growing up in shadowy Virginia mountains rather than in an urban world—it's equally dishonest to say that biology plays no role in who we are. We will never begin to understand ourselves if we insist on pretending that we are born as blank pages, that genetics provides no information as to whom we will be.

I agree with those baffled scientists on their main point. We have to look inside ourselves as well as out. I think we cheat ourselves and diminish ourselves by suggesting that we are made only of reactions to the world around us, as if we bring no inner self into our lives.

That doesn't mean that what happened to Carrie Buck is an acceptable risk. There's nothing acceptable about it. The worst possible end would be to learn nothing from what happened to her. Science has been used as an effective weapon against women, just as it's been used against a wide array of racial and ethnic groups.

Even today we have Richard Lynn arguing that biology explains why women don't perform as well as men on IQ tests, using the "credibility" of science to put women in their place. Clearly, his definition of that place is a lesser one than that held by men. There are lots of good political reasons for women to be very wary of trusting the pronouncements of research. The whole sorry eugenics episode should warn us against substituting political science for biological science. But that caution is different from refusing to allow honest research, or to acknowledge its results. Let's be vigilant; let's be cautious; let's even be suspicious. But let's also understand the science—its potential and its limits—so that we can prevent research from becoming trickery. To pretend that the facts don't exist will not protect us—and may harm us. By determinedly looking the other way, we may permit dishonest use of legitimate science to go unchallenged.

The risks, as illustrated by the eugenics movement, trouble those inside the science community as much as or more than they do those of us outside it. Since World War II, behavioral biology has continued, but in an unmistakably unfriendly environment. There have been furious protests, for instance, against researchers who explore the link between biology and violent behavior. And those who explore gender differences get accused of sexism for even taking an interest in the topic. Kim Wallen of Emory expresses some of the frustration: "What I really don't understand," he said, "is why there is such a negative reaction to what we know about sex differences. Maybe I have a perspective that reflects being a male, but, as I see it, there is a rough balance between the advantages and disadvantages that each sex experiences. What makes it so politically charged is that the advantages males have traditionally enjoyed allowed them to create a social system which devalued the advantages women have. So [for instance] the relatively constant interest in sex that males experience became the definition of sexual desire. Why isn't it the cyclic nature of female sexual desire, which is much more practical and ultimately critical to a functioning society? I think underneath all the conflict over gender differences is a tacit acceptance that male is 'normal.' "

As Wallen points out, there's nothing about primate biology that says male dominance is inevitable. There's nothing, even, that says that the bigger, bulkier sex takes over. The size difference may be an indicator of polygamy; it's not a straightforward prediction of social power. There are a number of primate species—rhesus macaques, squirrel monkeys—in which the difference in size between males and females is much larger than it is in humans, but the society is still a female-dominated one.

Wallen suggests that, in the end, it's not purely size and strength that allow a particular sex to achieve dominance. Power is built, he says, by relationships. "When I observe rhesus monkeys, I see these . . . striking differences which parallel many human sex differences. Yet the [monkey] society is almost completely controlled by females. Why is this? At some point, I think it is that rhesus females form strong female-female bonds and family bonds and this controls the males," Wallen says. "Stumptail macaques [a closely related species] have similar male-female sex differences. Both groups are polygamous; in both, the females raise the young. Yet the stumptail females don't show strong female-female bonds and the society is quite strongly controlled by despotic males. So, it seems to be that it is not whether the males and females differ in their behavior, but how the sexes form intrasexual relationships, which determine the character of the social structure."

Women form extremely tight small groups—essentially, our circles of friends. There's compelling evidence, of course, that women build stronger social networks, talk more easily about intimate matters, express emotions more surely. But if you take the feminist movement as an example, women seem less successful than men in putting together a network aimed at achieving power. Here, we have an organization, ostensibly formed solely to gain more power for women, and we have at the same time the "I'm not a feminist" reaction. A friend of mine, who has a high-level position in an influential corporation, once asked me why men seem to seek power for power's sake alone, while women don't. Is this a gender difference? Do we want power for different reasons than men? And does human biology predict that men will organize better in its quest? Those are fascinating and unanswered questions. Clearly, women haven't used their ability to build strong social networks in achieving power and dominance, as rhesus macaques have.

"Women are far more likely to 'nuke' a strong woman [than a man is]," reports psychologist Laurie Rudman of the University of Minnesota in Minneapolis. "It raises the question of whether women help keep other women down and the glass ceiling in place." Rudman did a study on that question, presented at the 1995 American Psychological Society meeting. She set up a series of job interviews in which women were asked to present themselves as either competent but self-effacing, or confident and assertive. Males and females were asked to interview the "applicant" and rate her in two ways. Did they like her? And would they hire her? Some were told they would benefit financially from hiring a more able woman.

If there was no personal gain involved, both men and women

preferred the meeker woman. But if the choice would earn them money, men chose the self-promoter. They also liked her more, in that case. Women still preferred the less assertive candidate, even if they rated Ms. Confident as more capable. Rudman was led to advise her audience of women to tailor their approach in interviews based on the sex of the interviewer: "If he's a man who really needs you, go ahead and self-promote. You might want to soften the edges if you're dealing with a woman."

Feminists themselves have bewailed the failure of women to bond in mutual support. University of California, Davis, anthropologist Sarah Hrdy points out that women historically have identified more with their families than with their overall sex. They've thus frequently sided with husbands or fathers in supporting "role definitions" that keep women out of politics and out of traditional male power positions. The failed Equal Rights Amendment, after all, was fought against as viciously by women as by men. Of course, there's a practicality to this—for many generations (until recently in this country) women have relied on men to support the household financially. In that context, building an alliance against one's provider would fit right into the old cliché of biting the hand that feeds you. For many, that might mean no food at all.

In the analysis of historian William O'Neill, the lack of a common goal has kept women from gaining the leverage that racial and ethnic groups receive from voting as one strong voice. In the context of power, why aren't women more united? Hrdy suggests that we will only understand that by understanding women in an evolutionary context, going back millions of years.

This returns me to the question: Can science explain the power difference between men and women? Is it written somewhere in our genes and their consequences? Is it woven into our evolutionary history? Are all the behaviors that fascinate me—from love to lust—simply part of an evolutionary arms race of sorts? Or do those questions ask too much of a basic biology and too little of us, denying the choices we make as conscious, complex beings?

If we accept that our society is at least slightly polygamous, then there is an evolutionary argument toward why males would seek power. Remember the chimpanzee societies studied by Frans de Waal, in which females formed "circles of friends" and males built a power-based hierarchy? The male winner—the one who got to be alpha male until other chimps staged a coup—was also the group's major stud, with primary access to receptive females. In terms of passing on his genes, power equaled reproductive success. De Waal suggests a potent ripple effect from what is, on one level, a fairly straightforward goal—to have more kids: "The unreliable Machiavellian nature of the male power games implies that every friend is a potential foe, and vice versa." Yet in order to survive, chimpanzee males also learn that alliances are essential; no male ever knows when he may need his strongest rival, de Waal points out. Coalition and power games are fundamental to male chimpanzee societies. Chimpanzee females, though, don't seek such political power; they seek social support. And they appear less pragmatic than males about rebuilding relationships; they are slower to forgive, and in the sense that it's more vulnerable to an emotional response, the alliance is not as strong.

If we think of chimpanzees as somewhat like our early ancestors, then it makes sense that we might still maintain similar behaviors. And that's what anthropologist Bobbi Low argues, that early human alliances were very comparable to those formed by chimpanzees. Women might have banded together to share information about child care, families, and household needs—and perhaps, in ancient times, foraging together for food. One might also predict that family itself would be important in such bonds, so that mother-daughter, sister-sister relationships would be especially strong. Males, on the other hand, might have been interested early in hierarchy, and in who would be on top. They'd thus form an alliance of people who would be most useful—relatives, friends, acquaintances—as long as it worked. This kind of history would predict that between father and son, brother and brother, there would be a greater capacity for emotional distance, for distrust.

In fact, it's not hard to find just those patterns in human society. Across cultures—the pastoral Mokugodo of Kenya, the Trinidadians, the Micronesian Islanders, the Mormons—status and wealth correlate with male reproductive success. Studies by Low in Sweden and Germany have found that richer men are more likely to marry younger women—raising the possibility of having more children—and to remarry if a marriage fails. In other words, the wealthy man, the one who controls the resources, has the most children, is the alpha male. Money equals power. In some societies, such as the Yanomamo Indians of South America, status is enough. The Yanomamo don't acquire riches. But the chieftain, and the other bosses of the tribe, are most attractive to women. Power alone turns out to be an aphrodisiac.

"So that for men, having power in and of itself can be reproductively rewarding," Low says. "Straightforward status can set a man apart reproductively. It doesn't work quite that way for women. In the few societies where women wield substantial public power, as opposed to informal influence, they show no clear reproductive gain from doing so." In fact, Low says, there's frequently a conflict between political power and direct reproductive gain for women. The conflict may be no more than one of time and energy—and certainly, in the past, men didn't have that problem. The traditional male politician had a traditional wife to raise the children.

But even today, the burdens of parenthood fall differently on husband and wife. So, in the purely genetic sense, the advantages of power seem tipped to the polygamous male in particular. Societies around the world bear it out. But did the shape of those societies, did our culture itself, arise from a natural reproductive biology?

At the most basic level, one could argue that natural selection would favor competitiveness and risk-taking and power acquisition for males over females. There's a great example of the pros and cons of risk-taking in another, clearly polygamous species, the elephant seals.

Eighty percent of the elephant seal males die without ever fathering a child. There's an obvious, tremendous incentive to risk it all for a one-night stand, a now-or-neverness about the males' existence. Biologists describe polygamous males as running almost eternally on a "live fast and die" track. There's no similar gain for females in this scenario. All an elephant seal female has to do is sit still and wait for the dominant male to waddle over and run off the crowd of losing courtiers. Her biggest risk is that she may be literally squashed to death during sex. She's wise to save her energy for pregnancy. Her own risk-taking impulses tend to focus on protecting her child. Elephant seal males won't be around to help with that.

Notice we're not arguing that elephant seals are just like humans or vice versa. The whole point, though, of arguing any connection between biology and behavior is this: We are, also, a living species on this planet. Unless we're going to argue that we were dumped here by space aliens and operate totally differently from every other species on Earth, we have to accept that as with every other species, biology plays a role in our behavior. It's not an "if," but a "how," and "how much." What makes us so interesting, in this analysis, is that we make biology so complicated. We can override our impulses, as a sexually frustrated elephant seal cannot. Or we can follow our impulses in some very nasty ways. It's difficult—dancing on the edge of impossible—to sort out whether our behavior is dragging our biology or the other way around.

Low uses warfare—indeed, an all-or-nothing risk—to explore how biology might influence a culture in which war is a man's game. "In nonhuman species, damaging aggression is surprisingly widespread," Low points out in an evolutionary analysis of war. Almost all battles are centered on reproduction—fights over mates, fights over resources to get mates. Many habitats can take on the appearance of a combat zone. Among red deer, an estimated 13 to 29 percent of adult males die from battle wounds sustained in fighting other male deer. But winners gain enormous reproductive benefits. In other words, the risk pays off.

"Evolutionary theory predicts that potentially lethal conflict will occur when the possible reproductive rewards (mates, status, resources for mates) are high; and that, within mammals, males will more often be in a position to gain than females," Low writes. Let's stick with the deer briefly. What would a doe gain from lethal combat? The bucks are all in dedicated pursuit anyway. A male, gathering up power/wealth/status, may also gather a larger number of mates, therefore casting his genes into a wider net. But the number of males a female can seduce has no reproductive gain whatsoever. Likewise, a woman, if she chose, could sleep with a different man every other hour and still could have no more than one successful pregnancy a year. What, then, would be the evolutionary advantage of gambling her life in a battle for power that would gain her greater sexual access? "Throughout evolutionary history, men have been able to gain reproductively by warring behavior; women have not," Low notes. "The difference is approach; men seek resources for direct reproductive gain, women, when they are involved, seek resources for themselves and their offspring."

The same argument, of course, has been applied to the different level of aggression between men and women. For instance, a recent study by Colorado anthropologists Craig Palmer and Christopher Tilley suggests that the formation of young male gangs, with all the associated viciousness and violence, may be driven, partly, by reproductive competition: sexual access to females. By tracking a sexually transmitted disease outbreak in Colorado Springs during the early 1990s, they found that gang members not only achieved a kind of power/money status within their community, they had an unusually high number of female partners, compared to the nongang members involved in the outbreak.

The trick, here, is not to reduce this argument down to idiocy. It doesn't mean that women want only to sit at home nursing the infant and smiling at the flowers. Neither does it predict that every man is a closet warrior, itching to strap on shield and sword. It says only that at a fundamental reproductive level, combat seems to offer more advantages to men than to women. Low admits that even this is not popular with some members of the feminist movement, who want to argue that women would be mirror images of aggressive males if only they weren't suppressed. "There's a difference between what we like and what we observe," she says. "Science isn't about what we wish for. It's about what we see. And the scary thing about evolutionary biology is that sometimes it helps us to see why things we want are hard to achieve."

But Low makes another—and, I think, really fascinating—point. In the arguments above, reproductive success is defined fairly simply by lots of healthy, long-lived (and reproductive) children. But today, children may actually have a better chance of survival if families are small but flush. The parents can afford good health care; they can provide a good education; they have the financial wherewithal to build a future. And that, in turn, means reproductive success might be built upon a woman working, helping to enrich the family resources. "What happens is that the emphasis changes, becomes more on how much you can invest in the kids, not how many kids you have. Women can be highly successful not by marrying early and having lots of children—which is our evolutionary history—but by being more like their husbands and collecting resources.

"We're in this really amazing, novel period," Low says. "We're looking at a human ecology that just didn't exist before. Your child's success may depend on whether he or she goes to college, even what college it is. So that we're in a situation where numbers aren't the answer and investment is. The most tangible currency is education. That's especially true in Western cultures like our own. We're creating this niche that hasn't existed—and what it means for changing roles of men and women is something we're all trying to figure out." Does that—in the United States or Europe, for instance—put biological pressure on an ancient power balance? Does it mean that it's in everyone's reproductive interests for women to move up the corporate ladder? If anything, this would be a pretty specific evolutionary niche. We certainly don't see it in Third World societies still pursuing the older, more-offspring-is-better strategies for evolutionary success. It's worth asking how long this approach can be successful, if resources continue to diminish. Some scientists theorize that women have gained the most power in countries where resources seem abundant and, therefore, sharing comes easier—at least for now.

Evolutionary biologist Judy Stamps reminded me that dissecting cultural evolution is an extremely difficult business, though. We can put together some compelling theories, analyzing animal behavior and modern human behavior, but there are caveats to both. No other species—not red deer, not elephant seals, not any of the primate societies I've referred to in these chapters—is a perfect model for our own. I really like the comparison to the power struggles of chimpanzees, but even that has to be handled with caution. If we simply accept a chimp model, how do we account for the multilayered influences of human culture?

There could well be cultural evolution in animals—we clearly have a history of underestimating our fellow species—but we don't yet know how to tease that out. And as for human behavior today, we don't know how much is clearly representative of the past and how much is fairly recent. Stamps makes the argument neatly in a further discussion of war:

"Human history suggests that making war and being aggressive, especially with ever more effective weapons, has been a good way to exterminate more passive cultures," Stamps says. "Look at what is happening now to Buddhist Tibet. Thus, what we have left today is what is left after all of this aggression that humans have engaged in over the past 4,000 years or so. However, with respect to genetic factors' influence on behavior, 4,000 years is nothing. Millions are more like it. It is also worth considering that before our brains got big enough to make effective weapons, humans were pretty deficient with respect to anything that could inflict serious damage on opponents. I can't see warfare being effective if opponents had to beat one another to death with their fists! Thus . . . what we see today around us is certainly a result of cultural evolution and selection, but I'm not sure what, if anything, it says about biology."

It's not as if, when scientists work in the field of evolutionary biology, they can go back in time and watch our ancestors at work and play. That's one reason that chimpanzees are so popular as a model, of course—because we believe that their behavior is similar to that of our very early ancestors. But even that conclusion is an educated guess. And, as Stamps so neatly points out, extrapolating from what we see today can be misleading. She could be right; for all we know the ancient gentle folk were wiped out by the first inventors of effective weaponry (sticks, rocks) and today's society is the surviving violent fringe element. Perhaps in the past, men and women had much more equal relationships as well. Again, there's room for argument and debate because we don't know. Adrienne Zihlman, at the University of California, Santa Cruz, has made a fascinating case for a female-centered early human society in a series of papers that builds a portrait of "woman the gatherer."

The time period that interests Zihlman begins some four to five million years ago, when the hominid line had just split away from the apes and our early ancestors were moving out of the forests and into the African savannah. Her theory is that these human ancestors—Australopithecines—were probably hunter-gatherers, with emphasis on the nuts and berries. An evolutionary half-step removed from chimpanzees, they probably were not organized or equipped to hunt large game, she suggests. Instead, males and females contributed to the food supply fairly equally. The parallel is direct to chimpanzees, whose diet is about 98 percent plants. In surviving hunter-gatherer human groups, between 50 and 90 percent of the diet is fruits and vegetables. "Meat was most likely not a major dietary component, although it was occasionally obtained in the manner of chimpanzees—by catching and killing small animals and pulling them apart with their bare hands.

"We do not envision a rigid division of labor by sex, but doubtless the frequency of certain activities varied both by sex and age," Zihlman says. "Females carrying young could not travel as far in search of food and would likely concentrate on reliable sources, whereas males might gather less consistently and more frequently bring back meat to share." In addition, she suggests there may have been early tools associated with gathering, similar to the sticks chimpanzees use to poke at termite nests. Or, perhaps, hominid mothers devised simple slings to carry their young after the human body became smoother and slicker, losing the thick hair and folds that allow young monkeys to clutch onto their mothers. It's possible, Zihlman proposes, that females thus were the first developers of simple tools, which, because they were made of organic material, have long since rotted away, unlike stone spearheads and knives.

It's also possible that early females were better at gathering up food than males; this could have raised their status in early hominid circles. There's no direct evidence of this idea, but Zihlman offers a fascinating comparison with chimpanzee behavior. For instance, chimpanzees on Africa's Ivory Coast have a passion for nuts and will carry them to rock outcroppings or big roots and then crack them open using smaller rocks or chunks of wood. This, indeed, is primitive tool use—comparable to the use of hammer and anvil. Adult female chimpanzees are a lot better at this than males and can get the toughest nuts open far more quickly. Why? Because males don't trust each other. They continually monitor the proximity and behavior of other group members and often stop cracking nuts in order to watch their companions. Females, far more relaxed in each other's company, cheerfully crack nuts whether or not they can see the nearest chimpanzee or not and, even if a conflict breaks out, keep cracking away. Unlike the males, they can thus get a lot more food from a single nut-cracking foray and have less need to travel farther in search of additional nutrition.

Zihlman theorizes that, essentially, home and hearth may have always helped bind people together. Among early humans, she believes the mother-child unit was most likely the "core unit" within the larger group, as with many other primates. Around this center were strong sibling bonds—a family. These small kin groups—Zihlman estimates their size at three to eight—might have shared food and helped to protect each other, especially the young and helpless. This community would have bettered the chances that the little ones would survive to adulthood, thus increasing the mother's reproductive success and, perhaps, selecting for such behaviors in the future. Says Zihlman, "The relatively strong bond we envision between mothers and sons and between male and female siblings would serve to integrate male hominids into the kin group, where they contributed to the survival of their shared genes through kin investment." In other words, woman is key to human evolution.

In fact, she thinks "man the hunter," the mighty warrior bringing down the great game species, probably didn't come about until much later, perhaps some half million years ago. Meat could have become more important as early humans migrated out of Africa, into the icier European climate, for instance, where plants are not available year-round. And, in Zihlman's scenario, it's possible that rigid division of sex roles didn't start until 10,000 or so years ago, when domestication of plants and animals really began and permanent settlements were built. Suddenly there was a need for someone to stay at home; there were home chores and field chores, defense of the settlement, and child care—the beginning, perhaps, of a true division of labor by sex.

Zihlman backs up her theory in a number of ways. For instance, so far no one has found stone hunting weapons that date back to the very early hominids, such as Australopithecus. This would tend to support the idea that those ancestors were not engaged in full-scale hunting and armed warfare. And she argues that the fossil evidence does not support extreme sexual dimorphism in our ancestors. That is, she believes that men and women have always been reasonably close in size. Men today average between 17 and 18 percent larger than women, nothing like the disparity among gorillas, in which males are nearly double the size of females. From studying monogamy, we know that species which form cooperative pair-bonds and share duties tend to be more similar in size. Zihlman's point, then, is that body shape suggests that humans, and their ancestors, have long had a cooperative relationship.

But the relationship between ancient bones and ancient bonds is not clear and remains a point of strong contention among anthropologists. African fossils have been found suggesting that early males were much larger; but there's also an ongoing argument over whether the bones have been dated correctly, and whether males have been correctly matched with females of their own time. The critical reminder in such disagreements is that these are theories—different people's reading of evidence based on scraps of bone and ancient bits of rock—and that there's room for both varied interpretation and mistakes.

I have to admit to a certain affection for an image of a woman-centered world. But if there was such a thing, what happened to it? What caused "woman the gatherer" to lose her status so rapidly? By the time Zihlman proposed her theory, in the early 1980s, the image we all had was of "man the hunter," traveling forth in all his power and glory, and woman waiting back at the campfire. Or as Zihlman said in a beautifully sarcastic speech to the American Association of Physical Anthropologists, "The men, of course, were the hunters, which presupposes a number of corollaries: men were making and using tools, obtaining and sharing food, forming male-male alliances, developing communication and becoming intelligent for more effective hunting. Pair bonds were the unit for sharing food. . . . There isn't much to elaborate on about the picture of women; it was drawn with disappearing ink. Home bases seemed to be invoked as places for early hominid women to stay home. While there, they were losing estrus, in order to be continually sexually receptive, thereby doing their part to reduce sexual chaos." Many female anthropologists have marveled that anyone could fall for this idea, considering that no primate society on Earth—from lemurs to chimps—includes homebound females.

Yet, not only did man-the-ultimate-hunter become the ruling theory, it became accepted as virtually unassailable, as truth. When Zihlman proposed her countertheory, it was—as she also remarked in the speech—"ignored, dismissed, and co-opted." The scientific literature continued to discuss man-the-hunter as if ancient women would never leave the campfire. Zihlman's theory was dismissed as biased, as if there was no previous bias in the other direction. Not that she was completely without influence. Other scientists adopted her emphasis on gathering and concluded that man was the primary forager. In exasperation, Zihlman remarked that "Beginning with the Greeks and continuing until the present day, only men are citizens, have jobs, argue about ideas. Taking this and projecting it into the past, we see that only men are breadwinners, with all that implies." The profession of anthropology itself, she said, operates within that essentially sexist framework. "I think sexism is more pervasive and more covert than racism, because assumptions about sex differences are largely unrecognized and unexamined. It seems all the more vital that physical anthropologists can and should transfer their evenhandedness and perspective developed in dealing with human variation to dealing with issues of sex and gender. The point is not whether there are differences but that the differences be approached and studied without value judgments."

Her concluding point, as you see, is the same as Wallen's: that we need to see gender differences without assigning them particular worth. Both imply that we've failed to do that. There's another idea of Wallen's that follows the notion that both men and women have certain strengths, but that men have been adept at using their strengths to put women at a disadvantage. If Zihlman is right, if we began as a female-centered population, how did we then become apparently male-centered, so that she herself ends up complaining about the blindness of sexism? What was the particular advantage that men employed and why? Was it simply the allocation of tasks as humans settled and built permanent homeplaces, that ended up giving men more control? Does it go beyond that? Does our biology necessarily include, perhaps even require, male-female competition to provoke new genetic development? And would that competition predict that either sex, if possible, would use some lucky advantages to become dominant?

Let's argue that men used a combination of larger size, muscles and corresponding strength; their inclination to network for power (as in the chimpanzee hierarchy strategies); and an aggressive "live fast and die" approach to life in order to gain species dominance.

They were able to do this because women did not build counter-power networks in the style of rhesus macaques. Over time, these advantages allowed men to set the reproductive rules—for instance, the sexual double standard—and that aided them more than it did women. Those rules, in turn, gave them an edge in the evolutionary arms race. Further, let's argue that holding a position of power helped develop a male biology that thrives on power. There are some terrific studies by Stanford University anthropologist Robert Sapolsky which show that the most dominant male baboons have a low level of resting cortisol—the hormone that rises so acutely with stress. But their system is unusually responsive to threat, with cortisol levels rising more rapidly than they do in monkeys with higher base levels. And since high cortisol stresses the immune system, the dominant animals tend to be healthier. We can play this both ways, of course: the male becomes dominant partly because of the advantage of being born with low-level cortisol, or the advantages of being dominant cause his cortisol to level off, reinforcing his position.

There's another angle to the advantage/disadvantage idea. Men might do more than play to their own strengths. They might also downplay any advantages that women have. For instance, the quicksilver responsiveness of testosterone might offer real male-male advantages in combat or competition. But it could also end up being a liability when it comes to attracting and negotiating a mutually satisfactory relationship with a female. If so, you might predict that dominant males, instead of controlling themselves, would use strength and aggression to control the females.

Kim Wallen puts it in this perspective: "The females, due to the cyclical nature of sexual interest, would themselves be more stable, more able to focus on work, more controlled than males. However, their effect on males could be exactly the opposite, making the males less stable by introducing sexual competition. Islam recognizes this effect of women, thus [the culture] hides them to reduce their impact on males. In contrast, we actually encourage women to dress in ways that men find sexually attractive. I have always wondered if this doesn't give women a competitive advantage in certain situations. I think it is why men sometimes fight to keep women out of certain situations. They are completely aware of the effect women have on them and know they need to avoid reacting to them on the basis of attraction. So it is easier to keep women out of the picture. There is a basis for a double standard. Whether women like it or not, they will have more of a physical impact on men than men will have on them. Males recognize this and, I think, resent it since it points out their inability to truly regulate their actions. Women also recognize this and can use and ignore it as they find appropriate. And they also resent the fact that men can't/don't treat them like [they do] other men."

But does this biological split make conflict inevitable? Isn't some of this reaction by men unnecessary? Couldn't women—if we're going to argue that they are more stable—make men more stable too, rather than more unstable? After all, scientists such as Alan Booth and Allan Mazur have found that men in steady, happy marriages show a drop in testosterone, a leveling off. I also raise this question with Wallen—whether the company of women might keep men more balanced in some ways. "One might argue—as you are suggesting—that a stable relationship, which eliminates sexual competition, might reduce a male's testosterone levels and increase his ability to focus on other things than sex," Wallen replies. "In other words, effective monogamy would be a female mechanism for reducing a male's competitive urge by reducing his testosterone levels."

And just at this moment, when I am beginning to think there's an actual, clear-cut answer in sight, Wallen adds that he's not sure he believes this mechanism actually works. "But it is interesting to speculate how the character of a monogamous relationship could alter the male's hormonal fluctuations and affect his interest in seeking other sexual partners."

Does living in quasi-monogamy thus affect our biology and make us long for full monogamy? As I suggested earlier, are we somewhere in between the two, perhaps traveling from polygamy toward monogamy? It's worth noting that polygamy predicts a division of labor, especially that males will not be involved in child care and that females, almost exclusively, will. Certainly, our culture has emphasized motherhood over fatherhood to the point that today's stay-at-home dads often complain of being excluded and undervalued. But is that labor division foretold by our biology? And if it is, does it mean that nurturing really does come easier to women overall? Can we change that? Do we want to?

Conversing about power with psychologist Sally Mendoza, I comment, "It seems to me I've got far more questions about this than I have answers." And she replies, "Well, then, you must be on the right track because that's what the field looks like, too." Not having all the answers is actually more fun—it allows us room to speculate, to indulge our curiosity, ask not only safe questions but ones that go right to the edge of improbability. There's also a risk in believing that knowledge is static, that science believes today what it did yesterday, and that the understanding and perceptions of research will ever arrive at some altar of total truth. Scientists, like the rest of us, figure it out as they go along, making mistakes and—with time—correcting them. The story of Carrie Buck should remind us of the danger in believing, at any given moment, that researchers can tell us precisely who anybody is or what that person is worth.

Consider today's arguments over the subtle differences in brain structure or the small overall difference in brain size. There are scientists who insist those variations are full of meaning, that they emphatically divide men from women, even that they mean men are mathematically more gifted, overall, than women. And there are scientists who will argue the same variations away completely, stating that they are totally meaningless and that even analysis is an indication of a sexist wish to put women down. We get the extremes because although we have the ability to find the differences, although high-end molecular biology and superimaging techniques will undoubtedly find more of them, we don't yet know how to interpret such data. And as a society, we're so tangled in webs of prejudice and preconception, subtle though they may be, that such test results are too easily snatched up as handy confirmations for cultural stereotypes—convenient props for one particular agenda or another. I don't mean to argue to the contrary that the findings are meaningless—anything but. But let us please acknowledge that we haven't fully understood them yet. We stand at a remarkable moment in behavioral biology when we finally know enough to ask some absolutely fascinating questions—and the answers are open-ended.

Gender biology has extraordinary promise if—and this may be an insurmountable if—we are willing to give it an objective hearing. It opens possibilities; it doesn't close them. And it doesn't segregate men from women. If anything, it does the opposite, emphasizing how intricately woven together we are in the design of evolution. We wrap around each other like different-colored threads in an astonishingly complex tapestry. Our whole history is about our mutual relationship and the way we've influenced each other. We may use hormones differently, but none of them—not testosterone, not estradiol, none of them—is exclusively a male or female province. Gender biology tells us that we are built of the same materials and that it takes only the barest slip or slide in organization to produce a man with breasts or a woman with masculinized genitals.

"Everything is biologically determined at one level," Wallen says, "but its expression is always an interaction with an environment. The failure of the deconstructionist view of gender was that it denied that the sexes were biased [predisposed] from the beginning. Apparently, to admit that males and females come with different predispositions was regarded as endorsing the worst kind of social discrimination. The boundaries between genders are naturally quite blurred. There are very few things that only males or only females can do. That, however, is not the same as saying there are few things that only males or only females are likely to do."

I would hope that gender biology could persuade all of us that a successful woman doesn't have to look like, or act like, a successful man. It tells us, as Wallen points out, that there is all manner of crossover potential. Nature can and does make a woman who's larger and stronger than a man, a man who's gentler than a woman, a woman who's edgy and competitive, and a man who's happy staying at home and raising the kids and creating a beautiful garden. If we allow it to be, biology is in some ways less limiting than the cultural stereotypes that we've imposed on each other.

We can make a convincing case that there's a biological origin in the way our cultures have assigned roles and attitudes to men and women. I think it's very possible that that explains, partly, why women seeking power have modeled themselves after men. The male paradigm stands in front of us: it so obviously works. Maybe networking and manipulation and brokering of deals comes more naturally to men, if we take a chimpanzee-model view of early evolution. Maybe that's why the feminist struggle has been so frustrating; maybe that's why when women consciously work to build "old girl" networks, they seem so closely to follow the "old boy" example.

But we have to get away from the outdated notion that biology assigns us a fixed place. If the study of gender biology—the whole sprawl of genes and everything else—tells us anything, it's that life is flexible, designed to adapt, tailored for change. And if, as Bill Rice's fruit fly work emphasizes, the environment we constantly adapt to includes each other, then perhaps women can begin changing the environment, and men will respond and adapt. In this sense, there's an enormously empowering message out of studying the sexes. Fifty years ago, the polling firm of Roper Starch Worldwide asked men and women which sex had an easier time of it. Both sexes said women. When the firm did the same survey in 1996, both said men. "If one's goal is to have a life of ease, then women have not made a whole lot of progress," said Bradford Roper, the vice president and research director. But the Roper survey asked another question as well: If you were to be born again, would you rather come back male or female? Fifty years ago, despite their "easy life," most women wanted to come back as men. There were no men who wanted to come back as women. In 1996, men still liked being men. But women wanted to stay female; in overwhelming numbers, they said that, if such a choice existed, they'd prefer to remain what they were.

If we needed the reminder, change is hard. Maybe we are pushing uphill against biology, to some extent. In a study of young college students, Sacramento psychologist Candace Adams found that both men and women expected to work when they graduated. But while 70 percent of the women also planned to do at least half of the household chores, only 49 percent of the men shared that expectation. Both sexes looked forward to leisure time and fun. But more than the men, the women also planned to spend a lot of time in personal relationships, talking to friends, raising children, nurturing all family members and, well, everything. In what is really a transitional time, when women both move in male domains and cling to female ones, it's no wonder that everyone perceives a woman's life as harder.

And yet, at some level, women obviously find more satisfaction in today's challenges than yesterday's more protected and confined lifestyle. There are even men, such as Warren Farrell, author of *The Myth of Male Power*, who argue that traditional roles have not served men so well. What's so great about being the sex predominantly slaughtered in war, for instance? What's so great about dying younger, withstanding disease with less resilience, and being less likely to survive a catastrophic trauma? And what's so great about an inflexible culture that demeans a man who wants to stay home with family?

I think Bobbi Low is right. In Western cultures, particularly, we live in a time of unprecedented change regarding the sexes. We've seen property laws change, voting laws change, access to status jobs change. It seems slow, maybe too slow. But if you think of such moves in terms of biology, the pace is extraordinarily rapid. Are we merely seeing the visible signs of an evolutionary gender-power shift, the kind studied by Bill Rice and Randy Thornhill? Are we seeing an environmental adjustment, as Low says is possible—that in a world of diminishing resources, the rules of reproductive survival change so that small families and a partnership to gather resources become the best choice?

And as gender roles change, will we change too? To some extent, biology predicts that we will. We know there's a complex exchange, that how we behave can change our biology. We may, indeed, further blur the lines between the sexes. Neurobiologist Marian Diamond may be right—that enough time spent with family and children will "gentle-down" the biology of a man. And sociologist Alan Booth may be right, too—that enough time pushed into competitive situations may be giving women a harder hormonal edge. Maybe one result of a more equally shared cultural environment is a more common behavioral biology. Perhaps we will build a self-reinforcing feedback loop, so that by acting more alike we will genuinely become more alike.

We don't understand all the fine details and interactions and byplays of behavioral biology. But we're trying. Partly because of our cultural concerns, we are driving science to try as well. To me, one of the more important questions is: Can we consciously choose where our biology will take us?

Since we do know that biology and behavior interact, can we make behavior decisions that will influence our biology? Can we consciously push society in ways that will allow women to compete more easily, or men to build stronger emotional bonds? And in doing so, are we also choosing to push ourselves, in terms of human evolution, further along those pathways? These considerations, in the realm of philosophical choice, may be what ultimately separates us from other species.

It's hard to imagine that we could agree to conduct ourselves along the lines of some great biological experiment. I'm not sure it's possible for any group to act so cohesively or so unselfishly in pursuit of a great mutual goal. And how do we agree on such a goal? How alike do we really want to be? As Wallen points out, science tells us that the lines between genders are blurred already. Do we want to erase them further, make them disappear? I don't—I'm not sure we even understand the consequences of pursuing that. Yes, I want the total partnership; I even think it would be fine if men catered to women the way titi males hover about their mates. But in terms of evolutionary consequences, would this mean that all men and women would eventually look alike? I'm not sure that it's total uniformity that we want. If we really dedicated ourselves to a kind of human monoculture, would we be nearly as interesting?

The most heartening part of seeing women rise in the power structure is not seeing them perform like powerful men, but like powerful women. My supervisor at the newspaper—a woman—successfully lobbied for a nursing lounge as part of a building remodeling, so that breast-feeding mothers now have a comfortable place to relax. The recent time period, when women's voices have really been heard, has seen a new emphasis on accommodating family needs, on day care, on flexible schedules, on the importance of a life outside the office. Maybe we still view this as much too female-loaded. It's *Working Mother* magazine, after all, that publishes the annual list of family-friendly companies. I'm still waiting—although not holding my breath—for *Working Father* to be published. Because, in the end, it's the fairness of partnership that concerns me, with room for each sex to choose, and even negotiate, what its part will be. In many ways it's a very good thing for women to be more comfortably competitive, for men to be more easily communicative, for each sex to learn and appreciate the strengths of the other.

The best sort of partnership—the most reinforcing, the most rational, and, I think, the most fun—will come only when we learn to fully appreciate and honor what we have in common while we continue to appreciate, and, yes, honor too, what makes us so confoundingly different.

NOTES

The information in this book is derived from sources that included personal interviews, research journals, scientific presentations, newspaper and wire service reports, and books.

Introduction

xiv–xv Comparing crime between men and women: "Gender, Race and Violent Crime," a presentation at the 1996 meeting of the American Association for the Advancement of Science by Roland Chilton of the University of Massachusetts, and Susan Datesman of Piper and Marbury, February 10, 1996; "Women and Violence: Victims, Offenders and Prevention Policies," a presentation at the 1996 meeting of the American Association for the Advancement of Science by Richard Rosenfeld of the University of Missouri at St. Louis, February 10, 1996. Robbery and assault figures from Anne Campbell, *Men, Women, and Aggression* (New York: Basic Books, 1993): 90–100.

xviii–xx Pioneer disasters and gender differences: Donald K. Grayson, "Differential Mortality and the Donner Party Disaster," *Evolutionary An-*

thropology 2 (1993): 151–159. Donald K. Grayson, "Human Mortality in a Natural Disaster: The Willie Handcart Company," *Journal of Anthropological Research* 52, no. 2 (Summer 1996): 185–205. Stephen McCurdy, "Epidemiology of Disaster," *Western Journal of Medicine*, April 11, 1994. McCurdy's quote is taken from a newspaper report by Deborah Blum, "Medical Students Learn from Donner Party Fate," *Sacramento Bee*, April 11, 1994, A-1.

xxii The cartoon I mentioned is by Edward Koren and ran in the *New Yorker*, August 19, 1996, 72.

Chapter One: *Splitting the Lark*

1–2 Poetry: George Gordon, Lord Byron. "So, We'll Go No More A-Roving," 1817; "She Walks in Beauty," *Hebrew Melodies*, 1815; from *Bartlett's Familiar Quotations* (Boston: Little, Brown and Company, 1980): 459–460. Emily Dickinson. "Split the Lark—and you'll find the Music," 1864; from *American Literary Masters*, vol. 1, ed. Charles R. Anderson (New York: Holt, Rinehart and Winston, Inc., 1965): 993–994.

2 John Maynard Smith's comments on the limits of science are from Nobel Conference XXIII, "The Evolution of Sex," held at Gustavus Adolphus College in 1987. Papers given at the conference were gathered in a book, *The Evolution of Sex*, edited by George Stevens and Robert Bellig (San Francisco: Harper & Row, 1988). Darwin's 1862 comment (page 3) on the mystery of sex is cited in the same publication.

3–4 David Crews' analysis of hermaphroditic chickens is from his "Constraints to Parthenogenesis," *The Differences Between the Sexes*, edited by Roger V. Short and Evan Balaban (Cambridge: Cambridge University Press, 1994): 23–52.

4 The timing of copulation is from Jared Diamond, "The Evolution of Human Sexuality," *The Third Chimpanzee* (New York: Harper Perennial, 1993): 75.

5–8 The section on the early origins of life, bacterial cell division, possibilities of cannibalism: From the Nobel Conference on the Evolution of Sex: John Maynard Smith, "The Evolution of Sex," 3–19; Lynn Margulis and Doron Sagan, "Sex: The Cannibalistic Legacy of Primordial Androgynes," 23–40; and William Donald Hamilton, "Sex and Disease," 65–99. The section on evolution of sex: John Archer and Barbara Lloyd, *Sex and Gender* (Cambridge: Cambridge University Press, 1985): 48–61.

8–9 The tale of the African snails and other examples of the importance of mixing male and female genes are detailed in Natalie Angier, "Male of the Species: Why Is He Needed," *New York Times*, May 17, 1994, Science Times.

9 The computer analysis of asexual reproduction appeared in the journal *Nature*, February 10, 1994, and was also reported in the Associated Press, February 9, 1994, by Malcolm Ritter in "Why Have Sex? Bad Genes and Parasites, New Study Suggests."

10 Russell Fernald, "Cichlids in Love: What a Fish's Social Caste Tells the Fish Brains About Sex," *The Sciences*, July/August 1993, 27–31.

11–14 Ursula Goodenough's work was presented at the 1995 meeting of the Council for the Advancement of Science Writing at Duke University in Raleigh, North Carolina. A synopsis titled "Sex Genes Across Species Evolving More Rapidly Than Others" also appeared on Quadnet, March 25, 1996, as released by Washington University in St. Louis.

13 The cultural history of incest is from Blair and Rita Justice, *The Broken Taboo* (New York: Human Science Press, Inc., 1979): 35–41.

14–16 The work on smell and immunology: Claus Wedekind, et al., "MHC-dependent mate preferences in humans," *Proceedings of the Royal Society of London (1995)*: 260, 245–249. See also F. Bryant Furlow, "The Smell of Love," *Psychology Today* 29, no. 2 (March/April 1996): 38–43. Also from proceedings of the International MHC meetings discussed on Sexnet in June 1996.

16 David Crews, "Courtship in Unisexual Lizards: A Model for Brain Evolution," *Scientific American* 255, no. 12 (December 1987): 116–121; and David Crews, "Animal Sexuality," *Scientific American* 270, no. 1 (January 1994): 109–114.

16–17 Information on gobies in Tina Adler, "Fishy Sex," *Science News* 148, no. 17 (October 21, 1995). Also from a presentation at the 1995 meeting of the Society for Neuroscience: Matthew Grober, "Serial Sex Reversal in Trimma Okinawae: A Model for Changing Sex in Changing Times," November 12, 1995.

17–18 Larry Shapiro's analysis of genetics and sex appears in "Threads of Life," *UCSF Magazine*, September 1992, as part of an article by Jeff Miller, titled "Sex, Chromosomes and Disease," 15–17.

20 Allen Enders's quote on God and miscarriage appeared in Deborah Blum, "The First Week—Pathway to Survival Holds Peril," *Sacramento Bee*, October 8, 1989, A-1.

22–23 Development of song and other sex differences in sound communication are discussed in two chapters of *The Differences Between the Sexes* (New York: Cambridge University Press, 1994). They are: Evan Balaban, "Sex differences in sounds and their causes," 243–272, and Manfred Gahr, "Brain structure causes and consequences of brain sex," 273–302. In his book on sexual orientation, *A Separate Creation* (New York: Hyperion, 1996): 91–97, Chandler Burr discusses both development of language in humans and song in birds.

23 Judy Stamps' discussion of genetic influences in katydids appears in her commentary, "Sociobiology: Its Evolution and Intellectual Descendants," *Politics and the Life Sciences* 14, no. 2 (August 1995): 191–193.

24 The roots of the words "androgen" and "estrogen" are given in Gail Vine's book, *Raging Hormones* (Berkeley: University of California Press, 1993): 18–19.

26 Birds, steroid hormones, and appearance are discussed in J. D. Reynolds and R. H. Harvey, "Sexual selection and the evolution of sex differences," *The Differences Between the Sexes* (New York: Cambridge University Press, 1994): 53–70.

26–28 On the Y chromosome and the disease vulnerabilities of men: Kenneth R. Miller, "Whither the Y," *Discover*, February 1995, 36–41; Jim Bull is quoted by Virginia Morrell, "The Rise and Fall of the Y Chromosome," *Science* 263 (January 14, 1994): 171–172. See also Richard L. Kelley and Mitzi I. Kuroda, "Equality for X Chromosomes," *Science* 270 (December 8, 1995): 1607–1610. On the sex-determining region of the Y chromosome, see Malcolm Gladwell, "Scientists Zero in on Key to Maleness: Single Gene May Trigger Sex," *Washington Post*, July 19, 1990.

28–30 Development of male and female embryos: *Sex and Gender: The Development of Sex Differences*, edited by Eleanor Maccoby (Stanford: Stanford University Press, 1966); Robert Pool, *Eve's Rib* (New York: Crown Books, 1994). Also, "Sexual Differentiation," a class lecture given by John Hearn, director of the Wisconsin Regional Primate Research Center.

30–31 Marc Breedlove's work on the bulbocavernosus is described in S. Marc Breedlove and David Crews, "Sexual Differentiation of the Brain and Behavior," *Behavioral Endocrinology*, edited by Jill B. Becker (Cambridge, Mass.: MIT Press, 1992).

31 Jared Diamond discusses the length of the male penis in *The Third Chimpanzee*, in the chapter "The Evolution of Human Sexuality."

32 A presentation of Leydig Cell Hypoplasia was made at the 1994 annual meeting of the American Society for Cell Biology by L. Laue, S. M. Wu, M. Kudo, A. J. W. Hsueh, J. E. Griffin, J. D. Wilson, D. B. Grant, A. C. Berry, G. B. Cutler, Jr., and W. Y. Chan, "The Gene Defect That Causes Genetically XY Males to Develop as Apparent Females."

32–34 Other genetic variations on XY are discussed in Marc Breedlove's chapter on sexual differentiation in *Behavioral Endocrinology*; see also his paper, "Sexual Differentiation of the Human Nervous System," *Annual Review of Psychology* 45 (1994): 389–418; also "Endocrine Abnormalities in Children," in Maccoby's *Sex and Gender: The Development of Sex Differences.*

Chapter Two: *Pinpointing the Difference*

37 For birdsong references, see chapter 1.

38 The Victorian male scientists' belief in the inferiority of female brains is set forth in detail in Cynthia Eagle Russett, *Sexual Science: The Victorian Construction of Womanhood* (Cambridge: Harvard University Press, 1989), particularly in the first two chapters, "How to Tell the Girls from the Boys" and "Up and Down the Phyletic Ladder," from which the background on Paul Broca and Havelock Ellis is drawn.

39, 57, 59 Scientists who believe that human intelligence is related to the individual's brain size: J. Philippe Rushton and C. Davison Ankney, "Brain Size and Cognitive Ability: Correlations with Age, Sex, Social Class and Race," *Psychonomic Bulletin and Review* 3 (1) (1996): 21–36. Also see Ankney, "Sex Differences in Relative Brain Size: The Mismeasure of Woman Too?," *Intelligence* 16 (1992): 39–336, and Richard Lynn, "Sex Differences in Intelligence and Brain Size: A Paradox Resolved," *Personal and Individual Differences* 17, no. 2 (1994): 257–271.

43 Raisman and Field's work is detailed in "The Brain As Sexual Organ," by Ann Gibbons, *Science* 253 (August 30, 1991): 957–960.

43–46 Differences in the hypothalamus, the Sexually Dimorphic Nucleus, and the other INAH cells are discussed in Breedlove's previously cited papers; in M. A. Hofman and D. F. Swaab, "The sexual dimorphic nucleus of the preoptic area in the human brain: a comparative morphology," *Journal of Anatomy* 164 (1989): 55–72; in D. F. Swaab,

L. J. G. Gooren, and M. A. Hofman, "The human hypothalamus in relation to gender and sexual orientation," *Progress in Brain Research* 93 (1992); and in Roger Gorski's presentation, "Structural Sex Differences in the Human Brain," at the 1996 AAAS Symposium on Sexual Differences in Brain and Behavior.

45 Simon LeVay's original report on INAH-3 cells was "A difference in hypothalamic structure between heterosexual and homosexual men," *Science* 253 (1991): 1034–1037. He elaborated further in his 1993 book, *The Sexual Brain* (Cambridge, Mass.: MIT Press). The *Discover* article on LeVay, which I cited, was by David Simmons, and appeared in the March 1994 issue, vol. 15, no. 3.

46 Michael Gazzaniga outlines his theories on brain differences and behavior in his book, *Nature's Mind: The Biological Roots of Thinking, Emotions, Sexuality, Language, and Intelligence* (New York: Basic Books, 1992).

47 The research cited from the Minneapolis Institute of Child Development was presented in "Gender Differences in Verbal and Spatial Functional Brain Asymmetries," by Dominick Wegesin on November 12, 1995, at the Society for Neuroscience annual meeting in San Diego, California.

47 The effects of DES are discussed in detail in the chapter "Echoes of the Womb" in Robert Pool's 1994 book *Eve's Rib* (New York: Crown Publishers, Inc.); also, see Breedlove's articles cited earlier.

47–48 The notion of a very different corpus callosum in men and women was described in two books: Jo Durden-Smith and Diane Desimone, *Sex and the Brain* (New York: Arbor House, 1983), and Anne Moir and David Jessel, *Brain Sex* (New York: Carol Publishing Group, 1991). Gorski also discussed differences in the anterior commissure and the corpus callosum at the 1996 AAAS symposium.

50–51 Joanne Berger-Sweeney described her work at an AAAS symposium, titled "The Developing Brain: Genes, Environment, and Behavior," on February 9, 1996. The finding that women did better on spatial tasks presented in a neutral way was by Matthew Sharps of California State University, Fresno, along with Jana L. Price and John K. Williams. The article, "Spatial Cognition: Gender," was published in *Psychology of Women Quarterly* 18 (September 1994): 413–425 and reported by the Associated Press, on November 28, 1994, as "Research Indicates Women Better at Spatial Tasks Than Thought." William Overman's comparison of abilities in children and monkeys was reported in "Little Humans and Little Monkeys Are a Lot Alike," by Bill Henrick, *Atlanta Journal and Constitution*, November 21, 1994, A-1.

51–53 Ruben and Raquel Gur have done numerous studies detailing the fine physiological differences between the brains of men and women, and I cite only a fraction of them. Among those: Ruben C. Gur, Raquel E. Gur, Walter D. Orbrist, Jean Pierre Hungerbuhler, Donald Younkin, Anna D. Rosen, Brett E. Skolnick, and Marin Reivich, "Sex and Handedness Differences in Cerebral Blood Flow During Rest and Cognitive Activity," *Science* 217 (August 13, 1982): 659–661; Roland J. Erwin, Maureen Mawhinney-Hee, Ruben C. Gur, and Raquel Gur, "Effects of Task and Gender on EEG Indices of Hemispheric Activation," *Neuropsychiatry, Neuropsychology, and Behavioral Neurology* 2, no. 4 (1989), 248–260. Raquel E. Gur and Ruben C. Gur, "Gender Differences in Regional Cerebral Blood Flow," *Schizophrenia Bulletin* 16, no. 2 (1990); Patricia E. Cowell, Bruce I. Turetsky, Ruben C. Gur, Robert I. Grossman, Derri L. Shtasel, and Raquel E. Gur, "Sex Differences in Aging of the Human Frontal and Temporal Lobes," *The Journal of Neuroscience* 14(8) (August 1994): 4748–4755; A. J. Saykin, R. C. Gur, R. E. Gur, D. L. Shtasel, K. A. Flannery, L. H. Mozley, B. L. Malmut, B. Watson, and P. D. Mozley, "Normative neuropsychological test performance: effects of age, education, gender, and ethnicity," *Applied Neuropsychology* 2 (1995): 79–88.

53–54 Research into evolutionary reasons why males and females might have different spatial skills is outlined in Steven J. C. Gaulin, "Does Evolutionary Theory Predict Sex Differences in the Brain?" *The Cognitive*

Neurosciences, edited by Michael Gazzaniga (Cambridge, Mass.: MIT Press, 1995).

56 Thomas Bever's studies of spatial navigation appear in the chapter "Different But Equal" in Robert Pool's *Eve's Rib.* Bever has also published some of his work in the chapter "The Logical and Extrinsic Sources of Modularity," *Modularity and Constraints in Language and Cognition*, edited by M. Gunnar and M. Maratsos (Hillsdale, N.J.: Lawrence Erlbaum and Associates, 1992).

56–57 Jacquelynne Eccles reported on her findings at the 1996 AAAS Symposium on Sexual Differences on Brain and Behavior, in a presentation titled, "Family Influences on Gendered Behaviors in School and on the Sports Field." She raises similar concerns in a chapter, "Bringing Young Women to Math and Science," in *Gender and Thought*, edited by Mary Crawford and Margaret Gentry (New York: Springer-Verlag, 1989). Cultural influences on girls' math and science abilities have been explored by a number of other concerned researchers, such as Janet Hyde, of the University of Wisconsin at Madison.

57–58 The gender gaps in math and verbal skills were reported in Larry V. Hedges and Amy Nowell, "Sex Differences in Central Tendency, Variability, and Numbers of High-Scoring Individuals," *Science* 269 (July 7, 1995): 4045, and covered by the Associated Press by Paul Recer in "Decades of Testing Show Boys Dominate Top and Bottom of Scale," July 6, 1995.

60 Differences in the planum temporale are cited in Roger Gorski's 1996 AAAS presentation and in a presentation at the 1993 meeting of the Society for Neuroscience entitled "Gender-Related Dimorphism of the Planum Temporale: An MRI Surface Rendering Investigation," by Jennifer Kylnych of the Clinical Brain Disorders Branch, National Institute of Mental Health.

60 Witelson's work was presented at the 1994 meeting of the Society for Neuroscience and described in "Do women top men in brain power?"

by Bill Hendrick, *Atlanta Journal and Constitution,* November 13, 1994, 6.

60–61 The Gurs' study of "resting" brains was reported in Ruben C. Gur, Lyn Harper Mozley, P. David Mozley, Susan M. Resnick, Joel S. Karp, Abass Alavi, Steven E. Arnold, and Raquel E. Gur, "Sex Differences in Regional Cerebral Glucose Metabolism During a Resting State," *Science* 267 (January 27, 1995): 528–531.

61–62 The Shaywitzes' report on gender and language differences in the brain appeared in B. A. Shaywitz, S. E. Shaywitz, K. R. Pugh, R. T. Constable, P. Skudlarski, R. K. Fulbright, R. A. Bronen, J. M. Fletcher, D. P. Shakweiler, L. Katz, and J. C. Gore, "Sex differences in the functional organization of the brain for language," *Nature* 373 (February 16, 1995): 607–608.

62–63 The study discussed by Steven Petersen is from Randy L. Buckner, Steven E. Petersen, and Marcus E. Raichle, "Activation of human prefrontal cortex across different speech production tasks and gender groups," presented at the 1994 Society for Neuroscience meeting.

Chapter Three: *Heart to Heart*

65–67 The information on mothers carrying babies on the left side of their bodies comes from a personal interview with Frans de Waal, and the issue is discussed in more depth in his book *Good Natured* (Cambridge: Harvard University Press, 1996). The book contains a section on gender emotional differences on pages 117–122. De Waal praised Martin Hoffman's work in our discussion, and he cites him in the book as well. Hoffman's 1978 paper, "Sex Differences in Empathy and Related Behaviors," *Psychological Bulletin* 84: 712–722, is considered a classic in the field.

67 Robert Pool discusses the sensory differences between men and women in *Eve's Rib.* Saul Schanberg's work on the importance of touch appeared in "The Experience of Touch: Research Points to a Critical

Role," by Daniel Goleman, *New York Times*, February 2, 1988, Science Times. See also S. M. Schanberg, T. Field, C. M. Kuhn, J. Bartolome, "Touch: A Biological Regulator of Growth and Development in the Neonate," *Verhaltenstherapie* 3 (Suppl.) (1993): 15; Saul M. Schanberg and Tiffany M. Field, "Sensory Deprivation Stress and Supplemental Stimulation in the Rape Pup and Preterm Human Neonate," *Child Development* 58 (1987): 1431–1447.

67 The University of Wisconsin study on touch is cited in the January 1994 issue of *Brainwork*, published by the Charles A. Dana Foundation, in an article titled "High on Hugging," by June Kinoshita.

67–68 The research on differences in vibrations within the ears was reported in "Quiet-Eared Women and the Men Born with Them," *Discover*, May 1994, 14–15.

68–69 Anne Fernald sets forth her mother-child communication work in a chapter titled "Human Maternal Vocalizations to Infants as Biologically Relevant Signals: An Evolutionary Perspective," in *The Adapted Mind: Evolutionary Psychology and the Generation of Culture*, edited by Jerome H. Barkow, Leda Cosmides, and John Tooby (New York: Oxford University Press, 1992): 391–428. The research connecting holding a baby on the left and emotional response is from "Why You Hold Baby on the Left," *Parenting* (November 1996): 13.

69 Margie Profet, "Pregnancy Sickness as Adaptation: A Deterrent to Maternal Ingestion of Teratogens," *The Adapted Mind*, eds. Jerome H. Barkow, Leda Cosmides, John Tooby (New York: Oxford University Press, 1993): 327–366.

70–74 Frans de Waal discusses chimpanzee networking in his book *Good Natured*; in "Sex Differences in the Formation of Coalitions Among Chimpanzees," *Ethology and Sociobiology* 5 (1984): 239–255; and in F. B. M. de Waal, (1993), "Sex differences in chimpanzee (and human) behavior: A matter of social values?" In: *The Origin of Values*, M. Hechter, L. Nadel and R. E. Michod (eds.) (New York: Aldine de Gruyter, 1993):

285–303. His look at grudge-holding in rhesus macaques appeared in "Coping with Social Tension: Sex Differences in the Effect of Food Provision to Small Rhesus Monkey Groups," *Animal Behavior* 32 (1984): 765–773.

72 Jane Goodall's description of chimpanzee warfare is in *Through a Window* (Boston: Houghton Mifflin Company, 1990): 98–103.

74 The Deborah Tannen anecdote cited is in *You Just Don't Understand* (New York: Ballantine Books, 1991): 24–25.

75 Robert Sapolsky's comments on primate gossip are from his article in the March–April 1995 issue of *The Sciences*; also reported by the Associated Press on February 27, 1995, in "Gorillas Gossip, Baboons Like to Watch."

75–76 Human sex differences in gossip were reported in Robin Western, "The Real Slant on Gossip," *Psychology Today* 29, no. 4 (July/August 1996).

76–77 The study of children playing doctor is cited in Tannen's book *You Just Don't Understand*, but is based on work by Jacqueline Sachs, reported in "Young Children's Language Use in Pretend Play," *Language, Gender and Sex in Comparative Perspective*, edited by Susan Philips, Susan Steele, and Christine Tanz (Cambridge, Mass.: Cambridge University Press, 1987): 178–188. Janet Lever's much-praised observations in 1972 of children at play were published later as "Sex differences in the games children play," *Social Problems* 23 (1976): 478–487.

77–78 Michael Milburn and S. D. Conrad, "The Politics of Denial," *Journal of Psychohistory* 23, no. 3 (Winter 1996): 238–251.

78 Theories of whether women's sensitivity to nonverbal expression is suggestive of a submissive approach are discussed in Judith A. Hall and Amy G. Halberstadt, "Subordination and Sensitivity to Non-Verbal

Cues," *Sex Roles* 31:3/4 (August 1994): 149–166. The authors do not find evidence for that theory.

78–79 Roland J. Erwin, Ruben C. Gur, Raquel E. Gur, Brett Skolnick, Maureen Mawhinney-Hee, and Joseph Smailis, "Facial Emotion Discrimination: 1. Task Construction and Behavioral Findings in Normal Subjects," *Psychiatry Research* 42 (1992): 231–240; and Frank Schneider, Ruben C. Gur, Raquel E. Gur, and Larry R. Muenz, "Standardized Mood Induction with Happy and Sad Facial Expressions," *Psychiatry Research* 51 (1994): 19–31.

80–81 The commentary on girls and women using emotions more effectively than boys and men is from Daniel Goleman, *Emotional Intelligence* (New York: Bantam Books, 1996): 130–133.

82–88 The background on the NICHD study on children in day care comes from Deborah Blum, "Boys May Be More Emotionally Fragile," *Sacramento Bee*, June 10, 1996, A-1. See also Jay Belsky and Michael Rovine, "Nonmaternal Care in the First Year of Life and the Security of Infant-Parent Attachment," *Child Development* 59 (1988): 157–167.

88 Geraldine Dawson's study of children of depressed mothers was reported at a June 1996 conference, "Brain Development in Young Children," held at the University of Chicago.

89–90 Nancy Bayley's work is discussed in "Boys May Be More Emotionally Fragile" (by Blum, *Sacramento Bee*, June 10, 1996), the Laura Allen quote in Michael D'Antonio, "The Fragile Sex," *Los Angeles Times Magazine*, December 4, 1994. Sally Mendoza's discussion of emotional connection is in "Morality Not Limited to Humans, More Scientists Believe," by Deborah Blum, *Sacramento Bee*, March 11, 1996, A-1.

90 Bruce McEwen's research into the effect of emotional distress on brain structure is detailed in Bruce S. McEwen, "Stressful Experience, Brain and Emotions: Developmental, Genetic and Hormonal Influences," in *The Cognitive Neurosciences*, Michael S. Gazzaniga, editor (Cambridge, Mass.: MIT Press, 1995): 117–136.

91–92 Robert Wright's discussion of Charles Darwin's grief over the loss of his daughter is in Wright's *The Moral Animal* (New York: Pantheon Books, 1994): 176–179.

Chapter Four: *Perfect Partners*

94 The statistics on the rareness of monogamy were provided by William Mason, professor of psychology at the California Regional Primate Research Center at the University of California at Davis, and cited in Deborah Blum, "Monogamy: Til death do us part is rare for mammals; what about us?" *Sacramento Bee*, October 16, 1995.

97–98 Charles Darwin focused on sexual selection in *The Descent of Man and Selection in Relation to Sex* (London: John Murray, 1868). The quotes are from those writings. His approach to sexual selection is also reviewed in Michael White and John Gribbin, *Darwin: A Life in Science* (New York: Dutton, 1995); and in Adrian Desmond and James Moore, *Darwin* (London: Michael Joseph, Publisher, 1991).

98 Marlene Zuk, at the University of California, Riverside, has done a series of studies demonstrating that secondary sex characteristics, such as a rooster's bright red comb, can both attract females and signal reproductive health. Her work is described in: Malte Andersson, *Sexual Selection* (Princeton, N.J.: Princeton University Press, 1994): 27–28, 74–77, 346–347.

98–99 Malte Andersson discusses his work and theories of sexual selection in his book *Sexual Selection* (Princeton, New Jersey: Princeton University Press, 1994). In an article called "Female Choice in Mating," *American Science* 80 (March–April 1992): 145, anthropologist Meredith Small discusses Andersson's widowbird study and its implications.

100–101 Height and other physical characteristics that make men attractive to women are discussed in a fascinating chapter by Bruce J. Ellis,

"The Evolution of Sexual Attraction: Evaluative Mechanisms in Women," in *The Adapted Mind*, 267–288. In the same book, David Buss's data on what makes women attractive, and on whether the sexes share any common interests, are detailed in a chapter called "Mate Preference Mechanisms: Consequences for Partner Choice and Intrasexual Competition," 249–266. Recent research on sexual selection and symmetry was also featured in Sharon Begley, "The Biology of Beauty," *Newsweek*, May 27, 1996. Other sources include: Elizabeth Pennisi, "Not Simply Symmetry," *Science News* 147, no. 3 (January 21, 1995): 46–47; Elizabeth Pennisi, "Imperfect Match," *Science News* 147, no. 4 (January 28, 1995): 60–61; Natalie Angier, "Why Birds and Bees, Too, Like Good Looks," *New York Times*, February 8, 1994, Science Times; Randy Thornhill and Steven W. Gangestad, "Human Fluctuating Asymmetry and Sexual Behavior," *Psychological Science* 5, no. 5 (September 1994): 297–302; Randy Thornhill and Steven W. Gangestad, "Human Facial Beauty," *Human Nature* 4, no. 3 (1993): 237–269; Matt Ridley, "Swallows and Scorpionflies Find Symmetry Is Beautiful," *Science* 237 (July 17, 1992): 327–328.

101–102 In *The Red Queen* (New York: Macmillan Publishing Company, 1994): 291–298, British science writer Matt Ridley discusses monogamy and sexual selection in great detail, including the waist-to-hip ratio research, the importance of facial features, and the attractiveness of youth.

102–103 Statistics on Down's syndrome and the problems of the aging egg are from Deborah Blum, "Why Do Birth Defects Increase As Moms Age?" *Sacramento Bee*, February 15, 1994, A-1.

104–105 The importance of lactation and the hormones associated with it were covered by Natalie Angier in "Family Values," *New York Times*, April 29, 1996, Science Times.

105–106 The quote from Yitzhak Koch on breast milk comes from Deborah Blum, "Is mother's milk key to child's growth, future?" *Sacramento Bee*, July 8, 1996, A-1.

107–109 Titi monkey bedtime rituals are discussed in Gustl Anzenberger, "Social Conflict in Two Monogamous New World Primates: Pairs and Bonds," *Primate Social Conflict*, edited by William A. Mason and Sally P. Mendoza (Albany, New York: State University of New York Press, 1993): 291–330. The behavior of titi monkeys is also profiled in Sally P. Mendoza and William A. Mason, "Parental division of labour and differentiation of attachments in a monogamous primate," *Animal Behavior* 34 (1986): 1336–1347. Mason compares the partnership of titi monkeys with a number of polygamous primates in "Primate Social Behavior: Pattern and Process," *Evolution of the Brain and Behavior in Vertebrates*, edited by R. B. Masterson et al. (Hillsdale, New Jersey: Lawrence Erlbaum Associates, 1976): 425–455. Matt Ridley weighs the evidence for and against human monogamy in *The Red Queen*, 175–178 and 206–207. So does Robert Wright in *The Moral Animal*, 54–57, 89–104. Both focus on the issues of fidelity over the issues of partnership.

109–110 Bobbi S. Low has published a lengthy series of articles exploring different aspects of human mating, marriage, and coalitions. Among them: "Sex, Coalitions and Politics in Preindustrial Societies," *Politics and the Life Sciences* 11, no. 1 (February 1992): 63–80; "Ecological correlates of human dispersal in nineteenth century Sweden," *Animal Behavior* 44 (1992): 677–693; "Reproductive Life in Nineteenth Century Sweden: An Evolutionary Perspective on Demographic Phenomena," *Ethology and Sociobiology* 12 (1991): 411–448; and "Human Sex Differences in a Behavioral Ecological Perspective," *Analyse & Kritik* (July, S., 1994): 38–67.

110–114 The citations for David Gubernick's research into the California mouse and monogamy include: David J. Gubernick, Kathryn A. Schneider, Lisa A. Jeannotte, "Individual differences in the mechanisms underlying the onset and maintenance of maternal behavior and the inhibition of infanticide in the monogamous biparental California mouse, *Peromyscus californicus*," *Behavioral Ecology and Sociobiology* 34 (1994): 225–231; David Gubernick, "Biparental Care and Male-Female Relations in Mammals," *Infanticide and Parental Care*, S. Parmigiana and F. S. von Saal, editors (Chur, Switzerland: Harwood Academic Publishers, 1994): 427–463; Gubernick, "Oxytocin Changes in Males Over the Reproductive Cycle in the Monogamous, Biparental California Mouse, *Peromyscus californicus*," *Hormones and Behavior* 29 (1995): 59–73; Gubernick, "Postpartum maintenance of paternal behavior in the biparental California mouse, *Peromyscus californicus*," *Animal Behavior* 37 (1989): 656–664; Gubernick, "A Neuroanatomical Correlate of Paternal and Maternal Behavior in the Biparental California Mouse," *Behavioral Neuroscience* 107, no. 1 (1993): 194–201; Gubernick, Dale R. Sengelaub, and Elizabeth J. Kurz, "A maternal chemosignal maintains paternal behavior in the biparental California mouse," *Animal Behavior* 39 (1990): 936–942; Gubernick, "Mechanisms of sexual fidelity in the monogamous California mouse," *Behavioral Ecology and Sociobiology* 32 (1993): 211–219; Gubernick and Randy J. Nelson, "Prolactin and Paternal Behavior in the Biparental California Mouse," *Hormones and Behavior* 23 (1989): 203–210; Gubernick, Sandra L. Wright, and Richard E. Brown, "The significance of father's presence for offspring survival in the monogamous California mouse," *Animal Behavior* 46 (1993): 539–546.

114–116 There is a vast literature on oxytocin and vasopressin, far more expansive than indicated in my fairly concise section on the two hormones. As the following citations will indicate, I did more comprehensive research on these hormones than is probably reflected in the chapter. For those who wish to pursue them further, I am providing the complete list: Thomas R. Insel, Stephanie Preston, and James T. Winslow, "Mating in the Monogamous Male: Behavioral Consequences," *Physiology and Behavior* 57, no. 4 (1995): 615–627; Thomas R. Insel and Terrence J. Hulihan, "A Gender-Specific Mechanism for Pair-Bonding: Oxytocin and Partner Preference Formation in Monogamous Voles," *Behavioral Neuroscience* 109, no. 4 (1995): 782–789; Thomas R. Insel, Zuo-Xin Wang, and Craig F. Ferris, "Patterns of Brain Vasopressin Receptor Distribution Associated with Social Organization in Microtine Rodents," *The Journal of Neuroscience* 14(9) (September 1994): 5381–5392; T. R. Insel, J. T. Winslow, Z.-X. Wang, L. Young, and T. J. Hulihan, "Oxytocin and the Molecular Basis of Monogamy," *Oxytocin*, edited by R. Ivell and J. Russell (New York: Plenum Press, 1995); James T. Winslow, Nick Hastings, C. Sue Carter, Carroll R. Harbaugh, and Thomas R. Insel, "A role for central vasopressin in pair-bonding in monogamous vole species," *Nature* 365

(October 7, 1993): 545–548; Lowell L. Getz and C. Sue Carter, "Prairie-Vole Partnerships," *American Scientist* 84, no. 1 (January-February 1996): 56–62; Craig F. Ferris, "The Rage of Innocents," *The Sciences* 36, no. 2 (March/April 1996): 22–26. Also from less technical sources: Jamie Talen, "Is Love Really Chemistry?" *Chicago Sun-Times*, October 13, 1993, 38; Jamie Talen, "Why Rodent Is True to Mate," *Newsday*, October 8, 1993, 1; Natalie Angier, "What Makes a Parent Put Up With It All?" *New York Times*, November 2, 1993, Science Times; Nancy Touchette, "Vole Mates: Vasopressin Keeps the Home Fires Burning," *The Journal of NIH Research* 8 (January 1994): 41–46; Scott LaFee, "Love (and Hormones): Could It Be That Chemistry Is at the Heart of All of Our Warm and Fuzzy Feelings?" *San Diego Union-Tribune*, February 14, 1996, E-1.

116–118 John Wingfield's chapter in *The Differences Between the Sexes*, titled "Hormone-behavior interactions and mating systems in male and female birds," 303–330, provides a detailed and clear overview of the field. A good discussion of infidelity among bird species can be found in Tim R. Birkhead, "Mechanisms of Sperm Competition in Birds," *American Scientist* 84 (May-June 1996): 254–262.

118–119 The research citations on budgie parents include: Judy Stamps, Anne Clark, Pat Arrowood, and Barbara Kus, "Parent-Offspring Conflict in Budgerigars," *Behavior* 94 (1985): 1–40; Judy Stamps, Anne Clark, Barbara Kus, and Pat Arrowood, "The Effects of Parent and Offspring Gender on Food Allocation in Budgerigars," *Behavior* 101 (1987): 177–199; Judy Stamps, Barbara Kus, Anne Clark, and Pat Arrowood, "Social relationships of fledgling budgerigars, *Melopsitticus undulatus*," *Animal Behavior* 40 (1990): 688–700; Judy Stamps, Anne Clark, Pat Arrowood, and Barbara Kus, "Begging Behavior in Budgerigars," *Ethology* 81 (1989): 177–192.

123–124 The importance of being a father: Paul Roberts, "Father's Time," *Psychology Today* 29, no. 3 (May/June 1996): 49–56. Roberts discusses the importance of rough-and-tumbling and teasing by dads as well as studies along those lines by Jay Belsky and Ross Parke.

124–125 The study on fathers spending time with infants was by Anat Ninio and Nurith Rinott, and reported in "Fathers' Involvement in Care of Their Infants and Their Attributions of Cognitive Competence to Infants," *Child Development* 59 (1988): 652–663.

Chapter Five: *The Second Date*

128 The numbers cited on same-sex attraction, actions, and thoughts come from Randall L. Sell, James A. Wells, and David Wypij, "The Prevalence of Homosexual Behavior and Attraction in the United States, the United Kingdom and France: Results of National Population-Based Samples," *Archives of Sexual Behavior* 24, no. 3 (1995). Michael Bailey found somewhat lower numbers in his surveys but informed me that either finding is acceptable, since "no one really knows what the precise numbers are." The estimate of five percent of the U.S. population as homosexual is cited by Chandler Burr, in *A Separate Creation.*

131 Stephen Jay Gould outlines and argues against the sociobiological explanation of homosexual-as-altruistic in an essay titled "So Cleverly Kind an Animal," in *Ever Since Darwin* (New York: W. W. Norton & Company, 1977): 260–267.

132–134 Bailey and Pillard's pioneering studies are explained in Chandler Burr's *A Separate Creation* and in his earlier article, "Homosexuality and Biology," *The Atlantic* 271, no. 3 (March 1993): 47–65. Dean Hamer also discusses the studies, and their influence on his work, in Hamer and Peter Copeland, *The Science of Desire* (New York: Simon & Schuster, 1994). The journal citations are: J. M. Bailey and R. C. Pillard, "A genetic study of male sexual orientation," *Archives of General Psychiatry* 48 (1991): 1089–1096; Bailey and Pillard, "A genetic study of male sexual orientation" (letter), *Archives of General Psychiatry* 50 (1993): 240–241; and Bailey, Pillard, M. C. Neale, and Y. Agyei, "Heritable factors influence sexual orientation in women," *Archives of General Psychiatry* 50 (1993): 217–223.

134–137 Hamer's original report on Xq28 was: D. H. Hamer, S. Hu, V. L. Magnuson, and A. M. L. Pattatucci, "A linkage between DNA markers on the X chromosome and male sexual orientation," *Science* 261 (1993): 321–327. *Time* magazine's coverage of his finding was headlined on the cover of the July 26, 1993 issue as "Born Gay: Science Finds a Genetic Link"; the story, titled "Born Gay?" by William A. Henry III, was on pages 36–39. The response of the gay community was covered by Natalie Angier, in a *New York Times* story titled "Gays Weigh Benefit of Emphasis on Genetics," July 18, 1993.

Following Hamer's first report, the study was critiqued in *Science* 262 (1993): 2063–2065, by Neil Risch, Elizabeth Squires Wheeler, and Bronya J. B. Keats, and defended by Hamer and his colleagues. Hamer also coauthored, with Simon LeVay, "Evidence for a Biological Influence in Male Homosexuality," *Scientific American*, May 1994, 20–25. His second report on the linkage was published as: Stella Hu, Angela M. L. Pattatucci, Chavis Patterson, Lin Li, David W. Fulker, Stacey S. Cherny, Leonid Kruglyak, and Dean H. Hamer, "Linkage between sexual orientation and chromosome Xq28 in males but not in females," *Nature Genetics* 11 (November 1995): 248–256.

138–139 The overall estimates of male homosexual behavior in other species can be found in A. Perkins and J. A. Fitzgerald, "Luteinizing Hormone, Testosterone, and Behavioral Response of Male-Oriented Rams to Estrous Ewes and Rams," *Journal of Animal Science* 70 (1992): 1787–1794. The article also discusses same-sex orientation in male rams, as does John A. Resko, Anne Perkins, Charles E. Roselli, James A. Fitzgerald, Jerome V. A. Choate, and Frederick Stormshak, "Endocrine Correlates of Partner Preference Behavior in Rams," *Biology of Reproduction* 55 (1996): 120–126. Experiments with sexual orientation in animals are reviewed by Elizabeth Adkins-Regan in "Sex hormones and sexual orientation in animals," *Psychobiology* 16 (1988): 335–347.

139–140 Donald Symons, *The Evolution of Human Sexuality* (New York: Oxford University Press, 1979). Symons discusses sexual orientation in a chapter titled "Test Cases: Hormones and Homosexuals."

141–144 Awareness of gender identity: Diane Poulin-Dubois, Lisa A. Serbin, Brenda Kenyon, and Alison Derbyshire, "Infants' Intermodel Knowledge About Gender," *Developmental Psychology* 30, no. 3 (May 1994): 436–442. Portions presented in April 1991 at meetings of the Society for Research in Child Development, Seattle, Washington. Daryl Bem discusses differences in boy-girl play in "Exotic Becomes Erotic: A Developmental Theory of Sexual Orientation," *Psychological Review* 103, no. 2 (1996): 320–335. Other sources include Deborah Tannen's *You Just Don't Understand*, 44, 152–181; Carol Gilligan, *In a Different Voice: Psychological Theory and Women's Development* (Cambridge: Harvard University Press, 1993): 9–11, 16, 172. The ages at which children tend to segregate by sex are discussed in "Separate ways: Toy choice is an early sign of gender gap," by Richard Saltus, *Boston Globe*, December 16, 1993. How parents respond to their children's requests for toys is discussed in an article by Claire Etaugh and Marsha B. Liss, "Home, School and Playroom: Training Grounds for Adult Gender Roles," *Sex Roles* 26, nos. 3/4 (1992): 129–144. The differences between young musicians were written up in the April 20, 1996, edition #2026 of *New Scientist*, by Alison Motluk, p. 7.

145 Fagot's toy work is in Saltus' *Boston Globe* piece. Play among rats and monkeys is described in *Eve's Rib* and many other publications, such as Jeanne Brooks-Gunn and Wendy Schemp Matthews, *He & She: How Children Develop Their Sex-Role Identity* (Englewood Cliffs, New Jersey: Prentice-Hall, Inc., 1979): 126–169.

148 Richard Greene, *The "Sissy Boy Syndrome" and the Development of Homosexuality* (New Haven, Connecticut: Yale University Press, 1987). Greene's influential work is also discussed in Bem's paper and in Burr's book.

148–149 The study cited on the minimal effect of parental sexual orientation is Susan Golombok and Fiona Tasker, "Do Parents Influence the Sexual Orientation of Their Children? Findings from a Longitudinal Study of Lesbian Families," *Developmental Psychology* 32, no. 1 (1996): 3–11.

150 The citation for the Kinsey Institute study is: A. P. Bell, M. S. Weinberg, and S. K. Hammersmith, *Sexual Preference: Its Development in Men and Women* (Bloomington: Indiana University Press, 1981). An appendix, containing data tables, was published under the same title later that year. These findings are very consistent. Other studies that reached similar conclusions are: J. Michael Bailey, Jennifer Nothnagel, and Marilyn Wolfe, "Retrospectively Measured Individual Differences in Childhood Sex-Typed Behavior Among Gay Men: Correspondence Between Self and Maternal Reports," *Archives of Sexual Behavior* 24, no. 6 (1995): 613–621; and Nathaniel McConoghy, Neil Buhrich, and Derrick Silove, "Opposite Sex-Linked Behaviors and Homosexual Feelings in the Predominantly Heterosexual Male Majority," *Archives of Sexual Behavior* 23, no. 5 (1995).

152 The abilities of gay and straight men and women are investigated in: J.A.Y. Hall and Doreen Kimura, "Sexual Orientation and Performance on Sexually Dimorphic Motor Tasks," *Archives of Sexual Behavior* 24, no. 4 (1995): 395–407.

154 Jerome Kagan, *Galen's Prophecy: Temperament in Human Nature* (New York: Basic Books, 1994), offers a great overview of his work on shyness and heredity.

155 Attitudes: Amber Arellano, "Men and Women Differ on Views Toward Gays," *Detroit Free Press*/Knight-Ridder Wire, April 5, 1995. Bem is writing a political postscript to "Exotic Becomes Erotic," in which he discusses the surveys exploring attitudes toward biology and homosexuality. The manuscript is still in draft form.

Chapter Six: *The Big T*

158–159 The history of the search for testosterone is detailed in John H. Hoberman and Charles E. Yesalis, "The History of Synthetic Testosterone," *Scientific American*, February 1995, 76–81. The enthusiastic response to its synthesis is recounted in Gail Vines's book, *Raging Hormones* (Berkeley: University of California Press, 1993).

160, 161, 165 Yesalis's perspective on anabolic steroids comes from that article and from: "Steroids and Sports: What Price Glory," by Skip Rozin, *BusinessWeek*, October 17, 1994, 176–177, and "Anabolic Shock: Use of steroids among young growing at alarming rate," by Doug Bedell, *Dallas Morning News*, December 10, 1995, B-1. The quote from Neil Carolan of Syracuse is also from Bedell's article.

160 Coverage of the health benefits of testosterone are from "Sagging Psyches," by Maureen Dowd in a January 24, 1996, editorial for the *New York Times* and were explored at length in a *Newsweek* cover story, "Testosterone: Attention, Aging Men," by Geoffrey Cowley, September 10, 1996, 68–77. The comments by Norm Mazer of Theratech, Inc. are from that article.

161 Recent coverage of testosterone, illustrating its image problem, includes: Linda Henry, "Hormones from Hell," *Muscle & Fitness* 55 (June 1995): 90–94; Beryl Lieff Benderly, "The Testosterone Excuse: It's Time to Talk about Men's Hormone Problem," *Glamour* 92 (March 1994): 184–185; Barbara Grizzuti Harrison, "Are Men Just Born to Be Mean?" *Mademoiselle* 95 (February 1989): 102.

167 On exertion: Alan Booth, Allan C. Mazur, and James M. Dabbs, Jr., "Endogenous testosterone and competition: the effect on fasting," *Steroids* 58 (August 1993): 348–350.

168 The numbers on the rise and fall of testosterone in rhesus macaques before and after fighting were provided by Kim Wallen of the Yerkes Regional Primate Research Center, at Emory University in Atlanta.

170–171 Competition: Allan Mazur, Alan Booth, and James M. Dabbs, Jr., "Testosterone and Chess Competition," *Social Psychology Quarterly* 55, no. 1 (1992): 70–77; Alan Booth, Greg Shelley, Allan Mazur, Gerry Tharp, and Roger Kittok, "Testosterone, and Winning and Losing in Human Competition," *Hormones and Behavior* 23 (1989): 556–571; on spectators, Curt Suplee, "Score One for the Hormone Team," *Washington Post*, October 2, 1995.

174 On relationships: Alan Booth and James M. Dabbs, Jr., "Testosterone and Men's Marriages," *Social Forces* 72(2) (December 1993): 463–477.

175–176 Testosterone and aggression: James M. Dabbs, Jr., Timothy S. Carr, Robert L. Frady, and Jasmin K. Riad, "Testosterone, Crime and Misbehavior among 692 Male Prison Inmates," *Journal of Personal and Individual Differences* 18, no. 5 (1995): 627–633. Robert Prentky, "The Neurochemistry and Neuroendocrinology of Sexual Aggression," *Aggression and Dangerousness*, edited by D. P. Farrington and J. Gunn (New York: John Wiley & Sons, Ltd., 1985): 7–55. Other studies along the same line include: Carolyn Tucker Halpern, Richard Udry, Benjamin Campbell, and Chirayath Suchindran, "Testosterone and Pubertal Development as Predictors of Sexual Activity: A Panel Analysis of Adolescent Males," *Psychosomatic Medicine* 55 (1993): 436–447, and Alan Booth and D. Wayne Osgood, "The Influence of Testosterone on Deviance in Adulthood: Assessing and Explaining the Relationship," *Criminology* 31, no. 1 (1993): 93–115.

176 The study of boat-bound physicians is cited in a paper by Allan Mazur and Alan Booth, "Testosterone and Dominance in Men." The citation is: W. Jeffcoate, N. Lincoln, C. Selby, and M. Herbert, "Correlation Between Anxiety and Serum Prolactin in Humans," *Journal of Psychosomatic Research* 29 (1986): 217–222.

177 James T. Winslow and Thomas R. Insel, "Social Status in Pairs of Male Squirrel Monkeys Determines the Behavioral Response to Central Oxytocin Administration," *Journal of Neuroscience* 11(7) (July 1991): 2032–2038.

179–180 The information on noradrenaline and serotonin is from Deborah Blum, "Natural born killers may be more than a movie title," *Sacramento Bee*, October 19, 1995. The comments from Bruce Perry also come from that article.

182 The study of female athletes and steroids is: Stephanie Van Goozen, Nico Frijda, and Nanne Van De Poll, "Anger and Aggression in Women: Influence of Sports Choice and Testosterone Administration," *Aggressive Behavior* 20 (1994): 213–222.

182 James Dabbs's story of a woman with high testosterone was quoted in Deborah Blum, "Disturbing Link Between Testosterone and Aggression," *Sacramento Bee*, July 30, 1990, A-11.

182–185 The studies on testosterone and women are: Francis Purifoy and Lambert Koopmans, "Androstenedione, Testosterone, and Free Testosterone Concentration in Women of Various Occupations," *Social Biology* 26 (1980): 179–188, also discussed in some detail in Anne Fausto-Sterling's book, *Myths of Gender* (New York: Basic Books, 1992): 129–130. Udry's study, "Androgen Effects on Women's Gendered Behaviour," was coauthored by Naomi M. Morris and Judith Kovenock, in the *Journal of Biosocial Science* 27 (1995): 359–368.

185–187 Kagan's studies on cortisol and behavior are detailed in his book, *Galen's Prophecy.* And the study by Alan Booth and James M. Dabbs, Jr., "Cortisol, Testosterone, and Competition among Women," was in 1996 an unpublished manuscript, under review.

Chapter Seven: *The Cycle Game*

189–190 Eighteenth century surgeries on women are described in Brant Wenegrat's *Illness and Power: Women's Mental Disorders and the Battle Between the Sexes* (New York: New York University Press, 1995).

190–191 Estrogen receptors and the rareness of survivable estrogen defects are discussed in: Kenneth S. Korach, "Insights from the Study of Animals Lacking Functional Estrogen Receptor," *Science* 266 (December 2, 1994): 1524–1526.

191–192 Both the description of estrogens and the later explanation of the menstrual cycle are based on a variety of sources, but some of the clearest information can be found in Geoffrey Redmond, *The Good News About Women's Hormones* (New York: Warner Books, 1995).

193 Natalie Angier, "Why Babies Are Born Facing Backward, Helpless and Chubby," *New York Times*, July 23, 1996, p. 11, includes background on physicians of the eighteenth and nineteenth centuries accidentally infecting women in labor.

193 Comparative data on male and female life expectancies: Charles Mann, "Women's Health Research Blossoms," *Science* 269 (August 11, 1995): 766–770; and James R. Carey and Pablo Liedo, "Sex Mortality Differentials and Selective Survival in Large Medfly Cohorts: Implications for Human Sex Mortality Differentials," *The Gerontologist* 35, no. 5 (1995): 588–596.

194 Immunology and sex hormones: Virginia Morrell, "Zeroing In on How Hormones Affect the Immune System," *Science* 269 (August 11, 1995): 773–775.

193–195 Data on estradiol's protective effects in heart disease come from: Trish Gura, "Estrogen: A Key Player in Heart Disease Among Women," *Science* 269 (August 11, 1995): 771–773; "Estrogen and Your Arteries," *Harvard Women's Health Watch* 2, no. 11 (July 1995): 6; "Estrogen and Longevity," *Harvard Women's Health Watch* III, no. 6 (February 1996): 7; "Estrogen has a quieting effect on the blood's clotting cells," *USDA Agricultural Research Service Quarterly Report*, October/December 1995, 4; "Shape-shifting hormones keep hearts healthy," *New Scientist*, May 18, 1996, 10; Jared Diamond, "Why Women Change," *Discover* 17, no. 7 (July 1996): 130–137.

197 A discussion of tubular breast formation is in Michael Thomas and Robert Rebar, "Delayed Puberty in Girls and Primary Amenorrhea," from a talk for medical students by Rebar, chairman of the Department of Obstetrics and Gynecology at the University of Cincinnati.

198–199 Malcolm Bilmoria and Monica Morrow, "The Woman at Increased Risk of Breast Cancer," *CA: A Cancer Journal for Clinicians* 45, no. 5 (September/October 1995): 266; Graham Colditz is quoted in Sarah Richardson, "His and Her Hormones," *Discover* 17, no. 1 (January 1996): 82. Statistics on heart disease in women are from "Estrogen: Key Player in Heart Disease Among Women" (*Science* 269) and for breast cancer, from the American Cancer Society's publication *Cancer Facts & Figures: 1995.*

200 The report on "Alzheimer's Link to Estrogen" was in *Science News* 174, no. 5 (February 4, 1995): 75.

201 The studies on rats, dendrites, and the estrus cycle are detailed in Sandra Ackerman, "Sex Hormones and the Brain," *Brainwork*, January 1994, 5; and Sarah Richardson, "The Brain-Boosting Sex Hormone," *Discover* 15, no. 4 (January 1996): 30–32.

203–206, 216 On estradiol and female behavior: Pamela L. Tannenbaum and Kim Wallen, "Hormonal modulation of sexual behavior and affiliation in rhesus monkeys," in press, Proceedings of the NYAS. Rosamarie Krug, Manuela Finn, Reinhard Pietrowsky, Horst-Lorenz Fehm, and Jan Born, "Jealousy, General Creativity, and Coping with Social Frustration During the Menstrual Cycle," *Archives of Sexual Behavior* 25, no. 2 (1996): 181–199. "Thinking on Sex Hormones," *Brainwork* 5, no. 2 (March/April 1995): 12, reports on Karen Berman's work. Uriel Halbreich's work was profiled in Kathleen Fackelmann, "Forever Smart: Does Estrogen Enhance Memory?" *Science News*, February 4, 1995, 74–75.

206–207 Kathryn Senior, "How oestrogen keeps the blues at bay," *New Scientist* 145, no. 39 (January 14, 1995): 18, covers George Fink's research.

207 The comparative numbers on estradiol and testosterone were provided by Kim Wallen, Emory University.

207–210 Susan Faludi describes the DSM decision on how to handle premenstrual syndrome and her unhappiness with that decision in her 1991 book, *Backlash* (New York: Crown Publishing Co., 1991): 357–362. The catalogue of symptoms is listed in Anne Campbell's book, *Men, Women, and Aggression*, and the history and influence of Katherina Dalton's book, *Once a Month*, 5th ed. (London: Fontana, 1991), is described in Gail Vines's book, *Raging Hormones*, as is the failure of progesterone therapy.

209 The effectiveness of Prozac in treating PMS was reported in the *New England Journal of Medicine*, June 7, 1995, and received additional coverage from the Associated Press in a story by Daniel Q. Haney, "A New Use for Prozac: Relieving Severe Premenstrual Syndrome," June 6, 1995.

211–215 The statistics and background on the rates of female-versus-male depression are from Myrna A. Weissman and Mark Olfson, "Depression in Women: Implications for Health Care Research," *Science* 269 (August 11, 1995): 799–801. Susan Nolen-Hoeksema reviewed the subject in her book, *Sex Differences in Depression* (Stanford: Stanford University Press, 1990). There is also an excellent review of the reasons for depression differences among teenagers in Susan Nolen-Hoeksema and Joan S. Girgus, "The Emergence of Gender Differences in Depression During Adolescence," *Psychological Bulletin* 115, no. 3 (1994): 424–443, which includes a discussion of differing academic expectations by the parents of boys and girls.

213 John Krentz's Internet study of preferred body shape was presented at the 1996 meeting of the American Psychological Association in San Francisco, and reported by Keay Davidson in "Forget the Skinny Ingenue, Men Prefer Real Women," *San Francisco Examiner*, July 1, 1996.

215 Other studies of adolescence and depression: Xiaojia Ge, Frederick O. Lorenz, Rand D. Conger, Glen H. Elder, Jr., and Ronald L. Simons, "Trajectories of Stressful Life Events and Depressive Symptoms During Adolescence," *Developmental Psychology*, no. 4 (1994): 467–483. See also

Regina C. Casper, Joseph Belanoff, and Daniel Offer, "Gender Differences, But No Racial Group Differences, in Self-Reported Psychiatric Symptoms in Adolescents," *Journal of the American Academy of Child Adolescent Psychiatry* 35(4) (1996): 500–508.

217–218 The study of recollection and grief was included in Sharon Begley, "Gray Matters," *Newsweek*, March 20, 1995.

218–219 Statistics on the changing rate of male depression are from Brant Wenegrat's book, *Illness and Power* (New York: New York University Press, 1995): 148.

Chapter Eight: *Counterstrikes*

220–221 The study on the *fruitless* gene: Lisa Ryner et al., *Cell* (December 13, 1996); Deborah Blum, "Male sexual behavior gene in spotted fruit flies," *Sacramento Bee* (December 13, 1996): A-1.

221–223 William R. Rice, "Sexually antagonistic male adaptation triggered by experimental arrest of female evolution," *Nature* 381 (May 16, 1996): 232–234. The underlying toxicity of fruit fly sperm was covered in "In Effort to Lessen Genetic Competition, Fruit Fly Sperm Can Be Deadly," Natalie Angier, *New York Times*, January 1, 1995, Science Times.

223–227 Theories on rape and evolution: Randy Thornhill and Nancy Wilmsen Thornhill, "The evolutionary psychology of men's coercive sexuality," *Behavioral and Brain Sciences* 15 (1992): 363–421 (including 28 commentaries); Randy Thornhill, "Is There Psychological Adaptation to Rape?" *Analyse & Kritik* 16 (1994): 68–85. Previous articles included: Nancy Wilmsen Thornhill and Randy Thornhill, "An Evolutionary Analysis of Psychological Pain Following Human Rape: IV. The Effect of the Nature of the Assault," *Journal of Comparative Psychology* 105, no. 3 (1991): 243–252; and Nancy Wilmsen Thornhill and Randy Thornhill, "An Evolutionary Analysis of Psychological Pain Following Rape: I. The

Effects of Victim's Age and Marital Status," *Ethology and Sociobiology* 11 (1990): 155–176.

228–229 Flirting in bars and differential responses to conversation are discussed in Daniel Goleman, "Scientists Study the Nuances of Human Courtship Rituals," *New York Times*, February 13, 1995. Donald Symons cites male/female response to conversations and discusses the different arousal reaction to nude photos and pornography in chapter 6, "Sexual Choice," in his book, *The Evolution of Human Sexuality.*

230 The different responses to sexual fantasy are cited in Daniel Goleman, "Reviewing Sex Fantasy Research, Study Finds It Blind to Women," *New York Times*, June 13, 1995.

230–231 The response of men to hearing audiotape accounts of rape is discussed in Thornhill and Thornhill's "The evolutionary psychology of men's coercive sexuality," in *Behavioral and Brain Science* and in Thornhill's "Is There a Psychological Adaptation to Rape?" in *Analyse & Kritik.* The citation for Malamuth's influential study is: N. Malamuth, M. Heim, and S. Feshback, "Sexual responsiveness of college students to rape depictions: Inhibitions and disinhibitory effects," *Journal of Personality and Social Psychology* 38 (1980): 399–408.

231–233 Symons discusses the dilemma of female orgasm in a chapter titled "The Female Orgasm: Adaptation or Artifact?" in *Evolution of Human Sexuality.* Other theories appear in Randy Thornhill, Steven W. Gangestad, and Randall Comer, "Human female orgasm and mate fluctuating asymmetry," *Animal Behavior* 50 (1995): 1601–1615. Orangutan counterstrategy evidence is reviewed in Lee Ellis's book, *Theories of Rape: Inquiries into the Causes of Sexual Aggression* (New York: Hemisphere Publishing Corporation, 1989): 48–50. More on orgasm: Susan Sprecher, Anita Barbee, and Pepper Schwartz, "Was It Good for You, Too?: Gender Differences in First Sexual Intercourse Experiences," *The Journal of Sex Research* 32, no. 1 (1995): 3–5. Meredith Small provides evidence for the occurrence of female orgasm in other species in *Female Choices*, 137–148.

234–235 Masturbation: Harold Leinberg, Mark J. Ketzer, and Debra Srebnik, "Gender Differences in Masturbation and the Relation of Masturbation Experience in Preadolescence and/or Early Adolescence to Sexual Behavior and Sexual Adjustment in Young Adulthood," *Archives of Sexual Behavior* 22, no. 2 (1993): 87–97; J. Linn Allen, "The M word: It's joked about, whispered about and worried about, but it's hardly ever talked about," *Chicago Tribune*, February 1, 1995. The Kinsey Institute studies on masturbation are detailed in Vines's book, *Raging Hormones*, 24.

235–236 The influence of oral contraceptives on MHC detection is discussed in the Wedekind paper cited in chapter 1. The study of the !Kung San tribes and the connection between estradiol and female desire are in Sarah Blaffer Hrdy, *The Woman That Never Evolved* (Cambridge: Harvard University Press, 1981): 131–159. The platonic attitudes of women on the pill have been studied by Rosamarie Krug, of Germany, who has done a series of studies exploring the relationships between women using oral contraceptives and their abilities and attitudes. These include: R. Krug, U. Stamm, R. Pietrowsky, H. L. Fehm, and J. Born, "Effects of the Menstrual Cycle on Creativity," *Psychoneuroendocrinology* 19 (1994): 21–31.

236–239 Ron Nadler's orangutan work is described in Sarah Blaffer Hrdy, "The Primate Origins of Human Sexuality," *The Evolution of Sex* (New York: Harper & Row, 1988): 101–138. See also Cathy Yarborough, "The Behavioral Impact of Natural Hormones and Hormonal Therapies Studies in Yerkes Monkeys," *Inside Yerkes*, Spring 1991, 9–13; Ronald D. Nadler, Jeremy F. Dahl, Welwood C. Collins, and Kenneth G. Gould, "Behavioral and Genital Sequelae of a Combined Oral Contraceptive in Chimpanzees," *Proceedings of the Second International NCRR Conference on Advances in Reproductive Research in Man and Animals*, edited by C. S. Bambra (Nairobi: NCRR, 1994); Ronald D. Nadler, Jeremy F. Dahl, Kenneth G. Gould, and Delwood C. Collins, "Effects of an Oral Contraceptive on Sexual Behavior of Chimpanzees," *Archives of Sexual Behavior* 22, no. 5 (1993); Ronald D. Nadler, "The Chimpanzee: A Useful Model for the Human in Research on Hormonal Contraception," *The*

Role of the Chimpanzee in Research, edited by G. Eder, E. Kaiser, and F. A. King (Basel, Switzerland: S. Karger, 1992): 56–67.

240–241 The Coolidge Effect is discussed by both Symons, *The Origins of Human Sexuality,* and Small, *Female Choices,* 179–180, who emphasizes female interest in variety. Small further discusses sexual novelty-seeking in females in "Promiscuity in Barbary Macaques," *American Journal of Primatology* 20 (1990): 267–282.

242–245 The sexual habits and background of hyenas is discussed in Deborah Blum, "Hyenas, Sex: No Laughing Matter," *Sacramento Bee,* December 18, 1995, A-1. Female sexuality in pigs and rats is reviewed in Kim Wallen's "The Evolution of Female Desire," in *Sexual Nature, Sexual Culture,* P. Abramson and S. Pinkerton, editors (Chicago: University of Chicago Press, 1995), 57–79.

245–247 Bonobo sex is outlined in Small's *Female Choices,* 48–49, 128–147, 172–181. See also Frans de Waal, "Tension regulation and nonreproductive functions in captive bonobos," *National Geographic Research* 3 (1987): 318–335.

248–251 The varied reasons behind rape: Robert A. Prentky and Raymond A. Knight, "Identifying Critical Dimensions for Discriminating Among Rapists," *Journal of Consulting and Clinical Psychology* 59, no. 5 (1991): 643–661; Neil M. Malamuth, Daniel Linz, Christopher L. Heavey, Gordon Barnes, and Michele Acker, "Using the Confluence Model of Sexual Aggression to Predict Men's Conflict with Women: A 10-Year Follow-Up Study," *Journal of Personality and Social Psychology* 69, no. 2 (1995). The history of attitudes toward rape is in Ellis's *Theories of Rape*; the comments from the California legislator are in Vern Bulloch, *Science in the Bedroom: A History of Sex Research* (New York: Basic Books, 1994): 245–248; the Cornell University e-mail is in Ellen Goodman, "This year's winners . . . are all losers," *Boston Globe,* reprinted in *Sacramento Bee,* August 26, 1996, C-2. Recent rape statistics are from Ellis, *Theories of Rape,* 3–7; male response to pornography also in Ellis, *Theories of Rape,* 12–14. The sexual response of males after being charged up by hearing an audiotape of a grisly killing is cited in Daryl Bem's "Exotic Becomes Erotic."

251–252 Female counterstrategies are surveyed by Natalie Angier, in "Is Homo Sapiens Just a Wolf in Men's Clothing?," *New York Times*, October 10, 1995, Science Times.

Chapter Nine: *Once Divided*

254 The complaint by Bertrand Russell comes from his book *Marriage and Morals* (New York: Bantam Books, 1959): 17. Other attitudes toward women: Margo Wilson and Martin Daly, "The Man Who Mistook His Wife for a Chattel," *The Adapted Mind* (New York: Oxford University Press, 1992): 289–326.

254–255 Power and domesticity: Marjorie E. Starrels, "Husband's Involvement in Female Gender-Typed Household Chores," *Sex Roles* 31, nos. 7/8 (1994): 473–490. Minimum wage statistics in Gilbert Chan, "On the Bottom Rung," *Sacramento Bee*, May 12, 1996, E-1.

255–256 Low's look at men, women, and political power is in Bobbi S. Low, "Sex, Coalitions, and Politics in Preindustrial Societies," *Politics and the Life Sciences* 11 (1) (February 1992): 63–80. See also: Bobbi S. Low, "An Evolutionary Perspective on War," *Behavior, Culture and Conflict in World Politics*, edited by W. Zimmerman and H. K. Jacobson (Ann Arbor: University of Michigan Press, 1993): 13–55; Bobbi S. Low, "Cross-Cultural Patterns in the Training of Children: An Evolutionary Perspective," *Journal of Comparative Psychology* 103, no. 4 (1989): 311–319; Bobbi S. Low and Joel T. Heinen, "Population, Resources, and Environment: Implications of Human Behavioral Ecology for Conservation," *Population and Environment* 15, no. 1 (September 1993): 7–40; and Bobbi S. Low, "Human Sex Differences in Behavioral Ecological Perspective," *Analyse & Kritik* 16, S. (1994): 38–67.

256–261 The history of the eugenics movement, the attitudes toward "feeblemindedness," and Carrie Buck from: Paul Lombardo, "Eugenical

Sterilization in Virginia: Aubrey Strode and the Case of *Buck v. Bell*," dissertation, University of Virginia, 1982. The statistics on sterilization in the U.S. are from: *Eugenic Sterilization*, Jonas Robitscher, editor (Springfield, Illinois: Charles C. Thomas, 1973).

263–264 Laurie Rudman presented her research on self-effacing women at the 1995 meeting of the American Psychological Society in New York City; it was reported in Marilyn Elias, "Self-promotion might hurt women in workplace," *USA Today*, June 30, 1995.

264 Sarah Blaffer Hrdy raises the problem of women and political bonds in her book, *The Woman That Never Evolved*, 130. She also cites William O'Neill's work, "Women in Politics," *Female Hierarchies*, edited by Lionel Tiger and Heather Fowler (Chicago: Beresford Book Services, 1978): 219.

268 Craig T. Palmer and Christopher F. Tilley, "Sexual Access to Females as a Motivation for Joining Gangs: An Evolutionary Approach," *The Journal of Sex Research* 32, 3 (1995): 213–217.

271–275 Adrienne Zihlman's work on "woman the gatherer" reported in: Adrienne L. Zihlman, "Did the Australopithecines Have a Division of Labor?" *The Archaeology of Gender*, edited by D. Walde and N. D. Willows (Calgary: University of Calgary Archaeological Association, 1991); Adrienne Zihlman and Nancy Tanner, "Gathering and the Hominid Adaptation," *Female Hierarchies*, edited by Lionel Tiger and Heather Fowler (Chicago: Beresford Book Service, 1978): 163–193; Nancy Tanner and Adrienne Zihlman, "Women in Evolution, Part I: Innovation and Selection in Human Origins," *Signs* 1, no. 3 (Spring 1976): 585–607; Adrienne L. Zihlman, "Women in Evolution, Part II: Subsistence and Social Organization Among Early Hominids," *Signs* 4, no. 1 (Autumn 1978): 4–20; Adrienne L. Zihlman, "Woman the Gatherer: The Role of Women in Early Hominid Evolution," *Gender and Anthropology*, edited by Sandra Morgen (Washington, D.C.: American Anthropological Association, 1989): 21–40; Zihlman, "Women as Shapers of the Human Adaptation," *Woman the Gatherer*, edited by Frances Dahlberg (New Haven: Yale University Press, 1981): 75–119. Chimpanzee gathering is discussed in: Adrienne L. Zihlman, "Gender: The View from Physical Anthropology," *The Archaeology of Gender*, 4–10. The speech in which she talks about the reception of her theory is "Sex, Sexes, and Sexism in Human Origins," given as the annual luncheon address at the April 1995 meeting of the American Association of Physical Anthropologists, and reprinted in the *Yearbook of Physical Anthropology* 30 (1987): 11–19.

276 *Primate Social Conflict*, William H. Mason and Sally P. Mendoza, eds. (Albany, NY: State University of New York Press, 1993).

280 The changing attitudes of women toward their place in the world is from "Easy Sex," *Psychology Today* 29, no. 5 (September/October 1996): 22.

281 Expectations in: Candace B. Adams, Margaret S. Steward, Thomas L. Morrison, and Lisa C. Farquahar, "Young Adults' Expectations About Sex-Roles in Midlife," *Psychological Reports* 69 (1991): 823–829.

281 Warren Farrell's book arguing that men are the all-powerful sex is *The Myth of Male Power: Why Men Are the Disposable Sex* (New York: Simon and Schuster, 1993).

INDEX

Index

Index

Index

Index

Index